The Energy Transition 2024

Topical Conference at the 2024 AIChE Spring Meeting and 20th Global Congress on Process Safety

New Orleans, Louisiana, USA
24 - 28 March 2024

ISBN: 979-8-3313-1714-0

Printed from e-media with permission by:

Curran Associates, Inc.
57 Morehouse Lane
Red Hook, NY 12571

CURRAN ASSOCIATES INC. proceedings.com

Some format issues inherent in the e-media version may also appear in this print version.

AIChE
120 Wall Street, FL 23
New York, NY 10005-4020

Phone: (800) 242-4363
Fax: (203) 775-5177

www.aiche.org

Additional copies of this publication are available from:

Curran Associates, Inc.
57 Morehouse Lane
Red Hook, NY 12571 USA
Phone: 845-758-0400
Fax: 845-758-2633
Email: curran@proceedings.com
Web: www.proceedings.com

TABLE OF CONTENTS

ANALYSIS FOR ENERGY TRANSITION

DESIGN FOR ENERGY TRANSITION

FEEDSTOCK EXPANSION FOR ENERGY TRANSITION

MODELING FOR ENERGY TRANSITION

NEW TECHNOLOGIES IN ENERGY TRANSITION

TRANSPORTATION CONSIDERATIONS IN ENERGY TRANSITION

MANAGEMENT OF CHANGE AND TECHNOLOGY IMPLEMENTATION IN ENERGY TRANSITION

MANAGEMENT OF CHANGE AND TECHNOLOGY IMPLEMENTATION IN ENERGY TRANSITION II

NEW TECHNOLOGIES IN ENERGY TRANSITION II

MODELING FOR ENERGY TRANSITION II

170a Low-carbon Methanol Production via Temperature-Pressure-Swing Reactive Capture and Conversion of CO_2: A Combined Techno-economic and Environmental Analysis

Eric Tan, Jonathan Martin, Anh To, Daniel Ruddy

Methanol is a versatile compound that can be used as both a fuel and a chemical intermediate. It is a colorless liquid with a mild alcoholic scent and a moderately high boiling point. Methanol can be used in various settings, including energy generation, transportation, and fuel cells. It can also be utilized as a fuel in vehicles, power plants, and fuel cells. Methanol is also a vital chemical intermediate that facilitates the production of various chemicals and materials, including formaldehyde, methyl esters, olefins, and dimethyl ether (DME). Its versatility makes it an essential component of diverse industries, contributing to energy production, transportation, and the synthesis of numerous chemicals and materials. The current total U.S. annual methanol capacity is about 9.4 million tons. Therefore, low-carbon methanol holds a key for industrial and transportation decarbonization. Low-carbon methanol can readily be produced via CO2 hydrogenation with renewable hydrogen, for example, using NREL's recently developed temperature-pressure-swing reactive capture and conversion (RCC) of CO2 process (Figure 1). The RCC process makes use of natural gas fired power plant flue gas CO2 as the feedstock for methanol production, which would otherwise be emitted to the atmosphere, thus enabling carbon circular economy and industrial decarbonization. Comparative techno-economic analysis (TEA) and life cycle assessment (LCA) will help assess the economic feasibility and environmental sustainability of the RCC process and compare it with other incumbent technologies, including the benchmark methanol synthesis from syngas derived from natural gas steam methane reforming with carbon capture and storage.

170b Life Cycle Analysis of Hydrogen Production from Biomass: A Comparative Study with Steam Methane Reforming (SMR) and Electrolysis Considering Government Incentives

Maryam Akbari, Shadi Darvish, Sudhanya Banerjee, Amit Goyal

Development of a low-carbon energy ecosystem requires extensive analysis of novel and emerging technologies, especially taking into account the effect of supporting policies on the progress of sustainable energies. Utilizing biomass as a sustainable and abundant source for clean fuel production can significantly reduce greenhouse gas (GHG) emissions. In this study conducted by DNV, the life cycle assessment (LCA) of hydrogen production from biomass resources was analyzed using the GREET Model and compared with Steam Methane Reforming (SMR), which is the dominant hydrogen production pathway in the U.S., and electrolysis, an emerging clean alternative to current fossil-based technologies.

GHG emissions of the life cycle stages for each pathway including feedstock production and preparation, transportation, and conversion as well as hydrogen compression and transport are assessed. Our assessment includes the identification of the most promising feedstocks, renewable electricity, production technologies, and delivery of hydrogen to off-takers. We evaluate the impact of government policies such as Inflation Reduction Act (IRA) tax credits on repositioning the attractiveness of these hydrogen production pathways and show the critical role of supportive strategies on the future of clean hydrogen production pathways.

This presentation aims to shed light on project-relevant aspects that help with increasing feasibility and is recommended for enterprises considering prospects for production, certification, and validation of low-carbon intensity hydrogen using DNV's harmonized practices to meet the requirement in every legislation.

170c Refinery integration and resource optimization modeling quantifies benefits of utilizing existing infrastructure

Michael Talmadge, Nicholas Carlson, Emily Newes, Abhijit Dutta, Ryan Davis, Ling Tao, Eric Tan, Kristi Moriarty

NREL's Economic, Sustainability, and Market Analysis (ESMA) team utilizes optimizable refinery models, resource allocation models, and process simulation-based debottlenecking analysis to quantify opportunities for biofuel and bioproduct pathways through integration with existing refining infrastructure. The goals of analysis are to model the transition of refineries to renewable and circular feedstocks and quantify benefits to costs of biofuel production and CO2 abatement through utilization of existing refineries and product logistical infrastructure. This presentation summarizes progress in refinery and logistics infrastructure analysis in support of multiple ESMA projects.

170d Decarbonizing the Chemical Industry: Assessing Current Transition Technology Options and Future Possibilities

Tim Hansen, Ignasi Palou-Rivera, Anita Bahe

Given significant energy demand and utilization by the process industries, particularly in chemical manufacturing, the identification of alternatives in energy production, delivery, and use are primary near-term considerations for decarbonizing manufacturing. Various roadmaps and studies on chemical industry decarbonization point to a select technological solutions and approaches towards this goal. The viability of these alternatives varies in terms of technology-readiness, market-readiness, and systems integration potential. To better understand current and future applicability of potential technologies for industry to prioritize, a data base of transition technologies was developed including over 150 specific technologies of interest to primary industry sponsors. These range from process and utility electrification to CO2 capture and utilization to electrochemistry. Each technology was assessed for its development stage, performance indicators, technology readiness levels for targeted applications, and other relevant metrics. Of equal importance barriers and opportunities impacting commercial viability were part of the analysis including market mechanisms, transition risks, economics, regulatory, and policy implications. Presenters will provide insights garnered from this work informing near-term deployment potential and longer-term needs and opportunities.

151a Design and Development of Novel Sustainable Technologies

Alexandria Dahl, Eloy Flores III, Hsiang Yee Hoekstra, Michael Hartmann

This discussion will provide SwRI's knowledge and experience that has been gained from the design, build-up and operations of such processes at various scales. At SwRI, we engage in the process development and improvement for a variety of sustainable fuels and chemicals from lab to pilot scales. The applied research, development, testing and evaluation methods undertaken help bridge the gap between fundamental science and commercialization. The technologies developed include but are not limited to hydrogen production, CCU, renewable fuels, chemicals and materials production and conventional thermo, photo and electro-chemical conversions. The systems have been developed to ensure process safety, enhanced conversion efficiencies, improved energy and mass transfer as well as overall emissions reductions and carbon neutrality.

151b SES Design: A System-Centric Approach to Decarbonizing Industrial Energy Systems

Julia Jimenez Romero, Adisa Azapagic, Robin Smith

This work introduces an innovative software tool designed to facilitate the sustainable transformation of energy systems within the process industry. With a central focus on decarbonization, the tool shifts its perspective from unit-centric to comprehensive system-centric, optimizing the entire energy system by considering a diverse portfolio of technologies and sources simultaneously, rather than in isolation. Notably, it factors in various elements, including the carbon footprint of technologies, energy grid dynamics, weather data, taxes, and energy tariffs, ensuring adaptation to the most optimal design.

An inherent capability of the tool is its ability, if enabled, to achieve the optimal allocation of temperature levels for energy integration across a site. It also demonstrates versatility in working with sites requiring power and heat at multiple temperatures.

The tool's adaptability extends to diverse scenarios, considering fluctuations in utility costs over time and varying technology expenses, ensuring robust results across a spectrum of operational contexts. Serving as a crucial guide in transitioning from current to future energy systems, it meticulously outlines trade-offs between economic and environmental targets. The software facilitates decision-making by constructing a cost-effective roadmap for industrial energy systems, a pivotal step towards achieving net-zero emissions.

In summary, this cutting-edge tool offers an objective, holistic, and adaptable approach to the design and operation of industrial energy systems, making substantial contributions to the decarbonization goals of the industrial sector.

Uncover the *Secret Life of Catalysts* with Accessible Reaction Engineering Tools: DRM and RWGS

Anna Lee Tonkovich, Tonkomo LLC, Gilbert, AZ
Eric Daymo, Tonkomo LLC, Gilbert, AZ
Akash Shirsath, Karlsruhe Institute of Technology, Karlsruhe, Germany
Matthias Hettel, Karlsruhe Institute of Technology, Karlsruhe, Germany
Olaf Deutschmann, Karlsruhe Institute of Technology, Karlsruhe, Germany
Cody Lockhart, HeatPath Solutions/Nexceris, Lewis Center, OH
Matt Seabaugh, HeatPath Solutions/Nexceris, Lewis Center, OH

Abstract

Catalytic reactors are important in many application areas, ranging from hydrogen, fuels, plastics, pharmaceuticals, food products, emission control, and countless others, with reactors and catalysts engineered to minimize unwanted by-products. Catalyst development often involves testing small scale experimental systems that have vastly different heat and mass transfer characteristics as compared to those of commercial-scale reactors. For example, lab-scale powder bed experiments may be minimally affected by internal mass transfer resistances, while narrow tube diameters may limit radial temperature variability inside the packed bed. Catalyst coatings on structured reactors such as monoliths may have a thermal profile that is difficult to measure during testing. Batch autoclave reactors reduce external transport limitations but reveal little about what happens inside a catalyst where concentrations may be very different from those in the bulk mixture. Given these experimental challenges, translating performance between reactor scales, or identifying catalyst performance problems is greatly enhanced by accessible reaction engineering tools, such as DETCHEM PBR and DETCHEM CHANNEL. Dry Reforming of Methane and Reverse Water Gas Shift examples are given, where these tools predict coke within catalyst structures. For the DRM example, SEM EDS analysis confirmed coke formation inside the interior of a catalyst pellet and not in the shell coating surrounding a highly conductive core (HeatPath™), as predicted by DETCHEM PBR. With RWGS, a 50 μm thick catalyst coating was found to suppress coking in an adiabatic monolith, partially due to the surprising side-benefit of methanation supplying sufficient heat to prevent carbon activity exceeding a value of one.

Introduction

The path to decarbonization will require many different technologies (Figure 1A), with catalyst selection and reactor design driving the performance of each of these unique processes. For each of the technologies outlined in Figure 1A (as well as the countless others not listed), a wide range of challenges and constraints must be addressed to have a successful design (Figure 1B). Oftentimes these challenges are not fully addressed, relying on guesses, heuristics, or experience. Other options require extensive experimentation but that can be time consuming and expensive. Another option is to simulate processes using computational fluid dynamics (CFD) or other tools, but oftentimes such codes are expensive or built-in features don't fully describe the physics of the system of interest. One effect in particular that is often insufficiently analyzed is the mass transfer resistance favoring side reactions that can occur inside porous catalyst structures.

Practical catalysts balance non-linear multi-step reactions with second-order diffusion effects. This interplay can reveal unexpected findings that mask laboratory observations or scale-up problems. Critical diffusion dimensions that affect side reactions depend on the reaction network complexity. These critical diffusion lengths that inhibit side reactions may be on the order of just a few µm, or as large as 1000+ µm. The "secret life of catalysts" occurs inside catalyst structures (e.g., *pellets, powders, and wall coated catalysts*), where transport is controlled by diffusion and where series or series-parallel reactions can dominate. Although this internal region physically contains most catalyst sites, complex reactions inside pore structures are generally poorly analyzed and rarely optimized. Side reactions that run amok inside a catalyst can be conflated with an intrinsic catalyst problem and the catalyst practitioner may move to the next formulation or condition. Reduced performance at scale can become accepted, instead of being understood and improved upon.

The physics of coupled reaction and diffusion is not new, described by (among others) Turing[1], Aris[2] and Fogler[3]. Every undergraduate chemical engineer learns the principles of internal diffusion limitations and a simple effectiveness factor analysis in their reaction engineering class.

The problem lies in the use of simplified analyses. Reality is often more complex and interesting than a simplified analysis might conclude. So why is this known effect of internal diffusional resistance coupled with reaction mostly ignored or oversimplified? Accessible reaction engineering tools are not routinely available nor are results generally expected to challenge the interpretation of laboratory data. Here, accessible means readily usable with little programming or customization needed for the chemical engineer more interested in the application than the theory. Without such reactor simulation tools, it is difficult to see what one doesn't expect to find.

DETCHEM[4] is a suite of accessible reaction engineering tools designed to help to understand what happens inside a catalyst, integrating:

- multi-step reaction kinetics, either from microkinetic models with perhaps hundreds of steps or apparent rates with a few steps, <u>coupled</u> with
- catalyst internal pore diffusion that may be molecular, Knudsen, or mixed, <u>coupled</u> with
- realistic heat transfer models, such as state-of-the art correlations for packed beds, structures, or open channels, to model the flow of heat that results from exterior heating/cooling along with endo- or exothermic heat transfer demands linked with the kinetic mechanisms.

DETCHEM PBR and DETCHEM CHANNEL from the DETCHEM software suite are used for the present work. DETCHEM PBR is a one-dimensional packed bed reactor simulator and DETCHEM CHANNEL is a single channel reactor simulator for structured reactors, such as monoliths and microchannels.

Two examples will be discussed that reveal the secret life of catalysts, and for each of these examples DETCHEM tools will be used to analyze chemistry inside of catalysts. In Section 2, we investigate a Dry Reforming of Methane (DRM) Nickel catalyst that was shown to concentrate CO inside the pellet. In the second example (Section 3), a modeling analysis for RWGS was completed to show that coke can form inside catalysts, even if the exterior of the catalyst is observed to be coke-free. The RWGS example shows that through modeling, an elegant solution can be found to allow operation in a regime without coke formation.

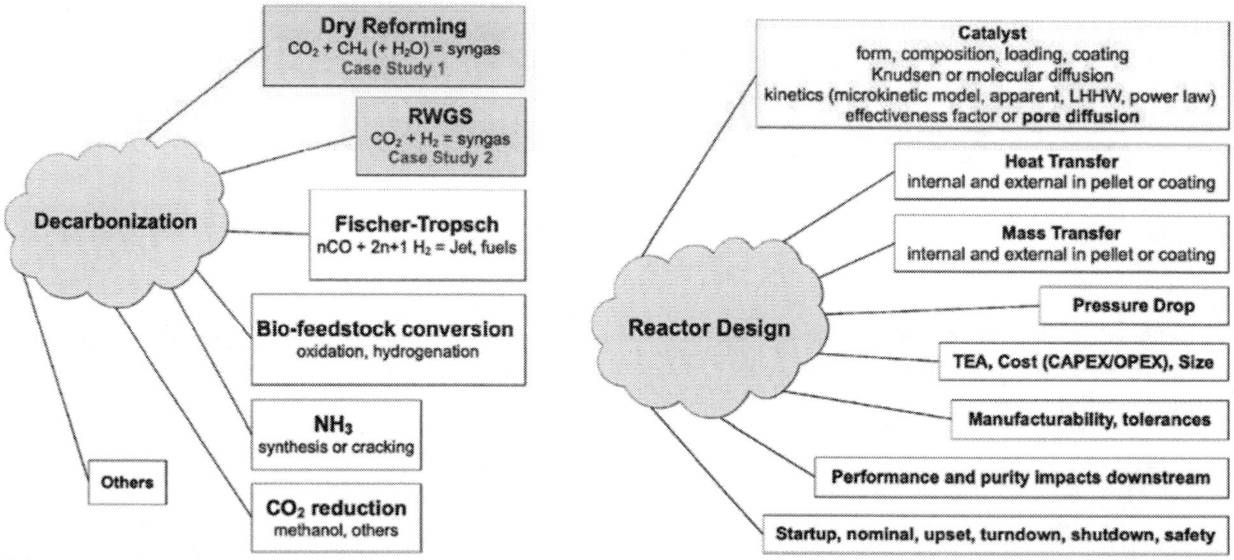

Figures 1A and 1B. A wide range of technologies (1A, left) and analyses (1B, right) are needed to address decarbonization.

Dry Reforming of Methane (DRM)

Summary experimental data for DRM collected with 6 mm Ni/alumina catalyst pellets are shown in Figure 2A. The catalyst deactivated when operated with a S:C below 1.5 and a high pressure of 20 barg but there was no visible external coke. DETCHEM PBR, however, predicted the formation of coke in the interior of the catalyst pellet for most of the catalyst bed (Figure 2B). This was confirmed by SEM EDS (Figure 3). Coke forms in the interior of the catalyst due to the concentrating effect of CO inside the catalyst pellet (Figure 4) which leads to the carbon activity exceeding one. When carbon activity exceeds a value of one, coke is thermodynamically favored to form. DRM modeling was based on an experimentally fit apparent kinetics for a simplified reaction network consisting of DRM, Water Gas Shift (WGS), and Steam Methane Reforming (SMR) steps.

One solution to avoid coke formation in DRM reactors is to use the Nexceris HeatPath™[5] catalyst, whereby the catalyst is coated on a dense, highly conductive core (Figure 5). At a steam to carbon ratio (S:C) down to 0.3 the catalyst ran for 450 hrs without coke formation (predicted by DETCHEM PBR modeling, and confirmed by SEM EDS of selected pellets, Figures 6A and 6B). Despite the lower volumetric catalyst loading of the 12wt.% Ni doped Al_2O_3-CaO-MgO material when applied as a thin coating on the HeatPath™ catalyst particle, the performance improved (Table 1) due to the elimination of side reactions inside the catalyst pellet. Without the proper accessible tools, such as DETCHEM PBR, finding an engineering solution would be difficult.

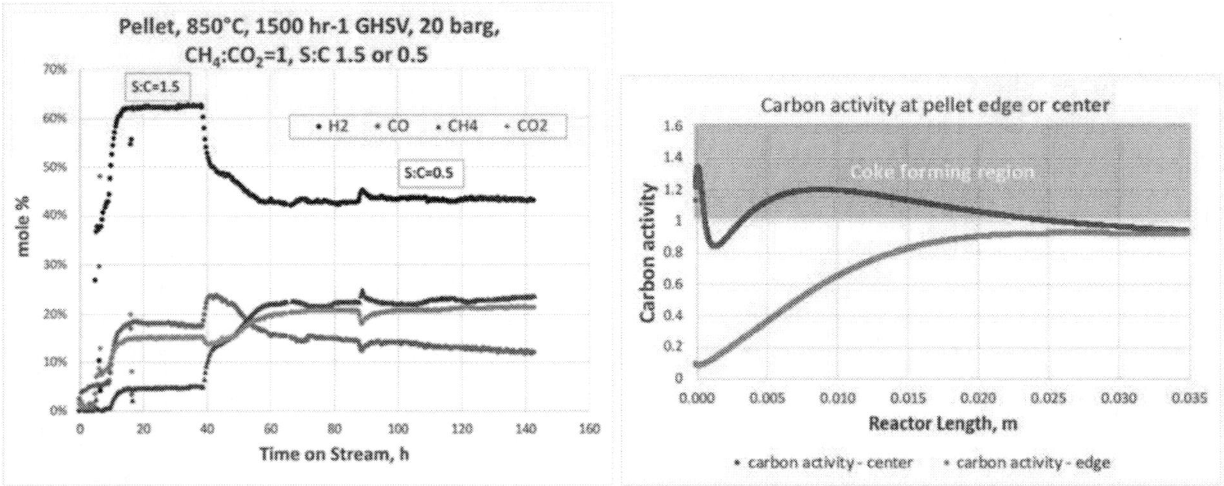

Figures 2A and 2B. DRM data (2A, left) and a DETCHEM PBR modeling prediction that coke under these experimental conditions should form inside the catalyst pellets, not on the exterior of the pellets (2B, right).

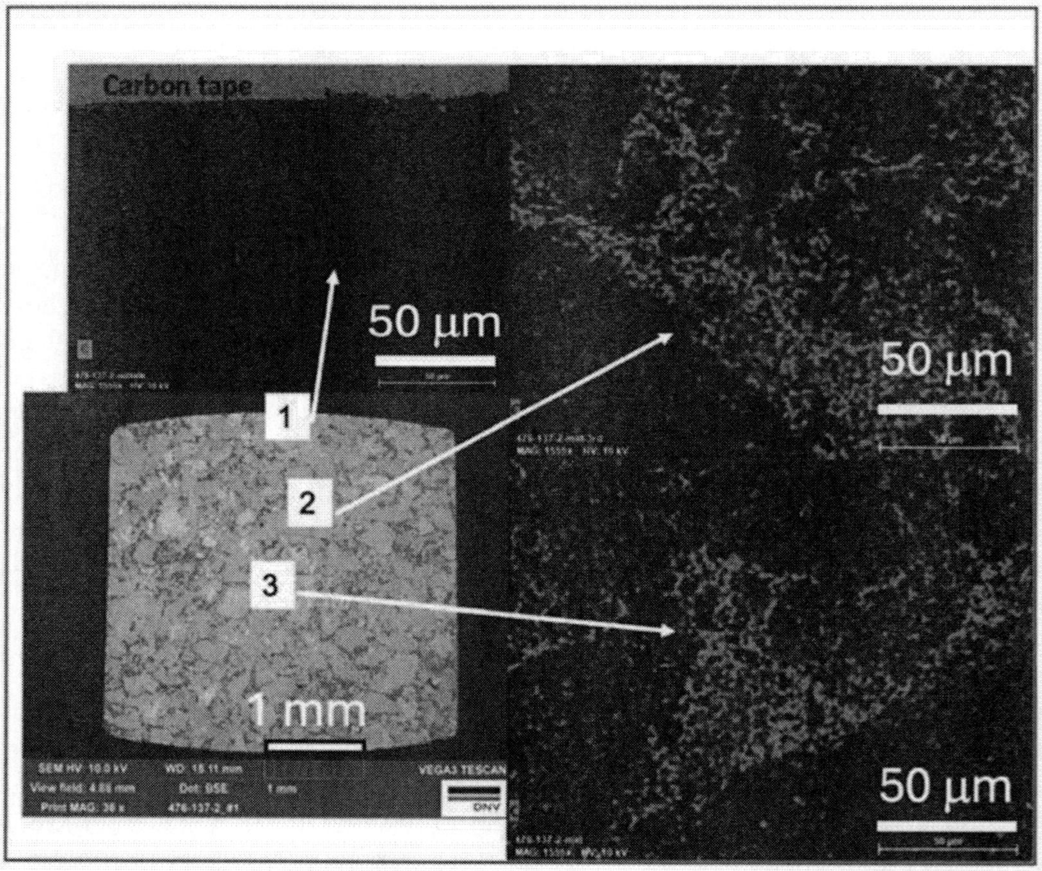

Figure 3. SEM analysis confirms the presence of coke inside, but not on the outside of the catalyst pellet, as predicted by DETCHEM PBR. The length scales have been enhanced for clarity.

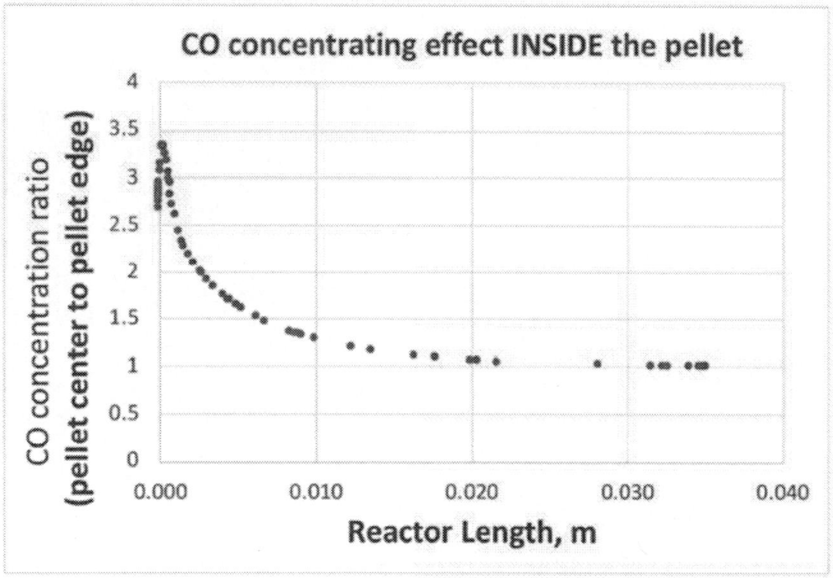

Figure 4. Coke forms due to a concentrating effect of CO inside the pellet, as predicted by DETCHEM PBR.

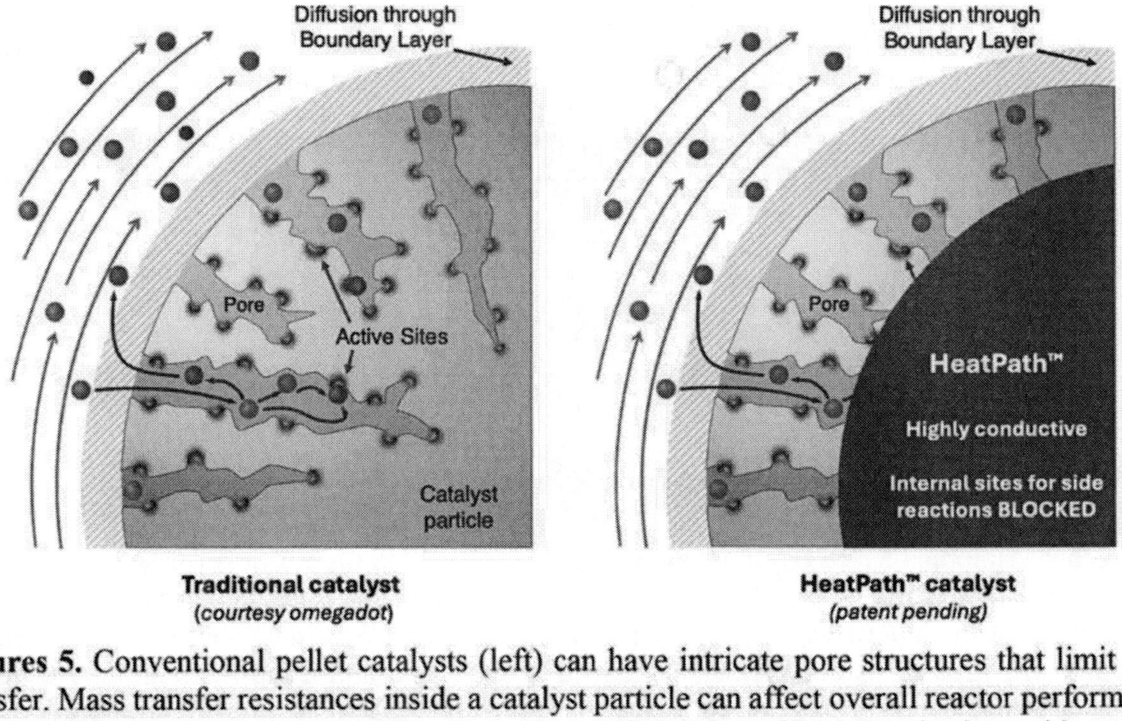

Figures 5. Conventional pellet catalysts (left) can have intricate pore structures that limit mass transfer. Mass transfer resistances inside a catalyst particle can affect overall reactor performance, but the effect of stagnant zones within the pellet can be mitigated by instead using a thin catalyst coating. Thin shell coatings on highly conductive catalyst cores are possible with Nexceris HeatPath™ (right).

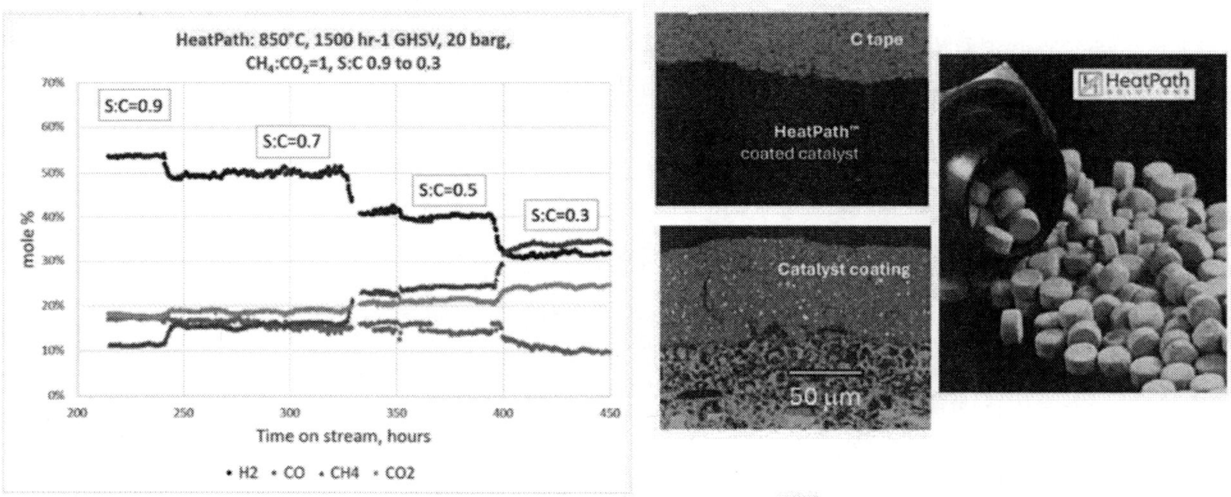

Figures 6A and 6B. DRM tests were repeated with HeatPath™ (6A, left). With the thin shell coating, coke does not form inside the catalyst (6B, right). The presence of carbon in the top of the SEM EDS dot map (orange color) is due to carbon tape, not coke.

Table 1. Summary of DRM results (GHSV=1500 hr^{-1}, P=20 barg, CH_4:CO_2=1, T=850°C). Despite the thin coating, the HeatPath™ pellet with a shell catalyst coating was more productive than a conventional pellet because the interior of the conventional pellet is where CO concentrated and unwanted side reactions – like coking – occurred.

System	S:C	g/g/h (CH_4)	g/g/h (CO_2)
Pellet	1.5	0.15	0.12
Pellet (*pre-coking*)	**0.5**	0.11	0.37
HeatPath™	1.5	5.88	4.38
HeatPath™	**0.5**	2.95	10.12

Reverse Water Gas Shift (RWGS)

In another example, a DETCHEM PBR analysis of RWGS on a Nickel-based catalyst was completed to determine how to best configure a catalyst to operate at pressure without coke formation. A microkinetic model with 6 gas phase species, 13 surface species, and 52 surface reactions was used to model the catalyst[6]. Both packed bed and a monolith structured reactor designs are considered in this design study. A packed bed reactor with 5-mm diameter catalyst particles exhibits a strong internal pellet CO concentrating effect and a carbon activity greater than 1 in a 0.5 m long, 4 cm diameter tube. As a result, the carbon activity at the center of the pellet is substantially higher than at the edge of the pellet and internal coking is predicted (Figure 7A).

To evaluate the effect of thin coatings and the potential for coke formation, the same RWGS microkinetics were applied to a model of a wall coated adiabatic monolith reactor using DETCHEM CHANNEL (Figure 7B). Catalyst thickness varied between 10 and 100 microns, with results presented here for two catalyst thicknesses: 50 and 75 μm. A critical catalyst thickness was identified for the structured catalyst, beyond which the carbon activity inside the wall coating can also exceed unity. Surprisingly, the detailed microkinetic model shows a small methane formation rate (selectivity <2%, Figure 8A) for which the highly exothermic methanation reaction can in some cases achieve a sufficiently high internal catalyst temperature (Figure 8B) to maintain a local carbon activity less than one, overcoming carbon formation in the monolith. An alternative catalyst formulation that suppresses methane formation would not benefit from this extra heat and would be more likely to form coke in the slightly colder catalyst interior. A highly conductive coated catalyst pellet could also overcome the predicted carbon formation.

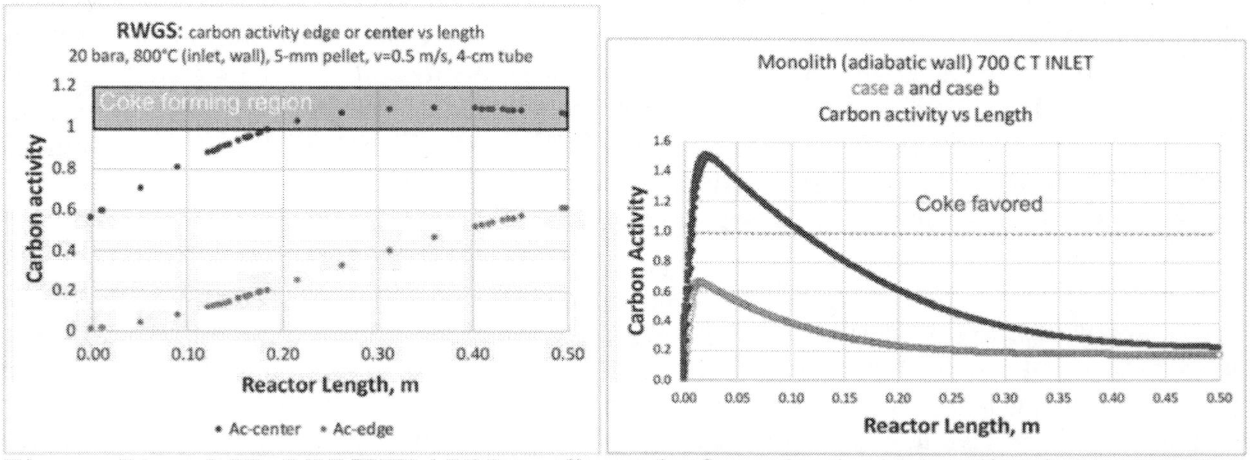

Figures 7A and 7B. DETCHEM PBR predicts coke formation in the interior of catalyst pellets (7A, left), as does DETCHEM CHANNEL for 75 μm thick (case b) coatings (7B, right) in a washcoated monolith. Coke is not predicted, however, for 50 μm thick (case a) washcoated coatings in a monolith.

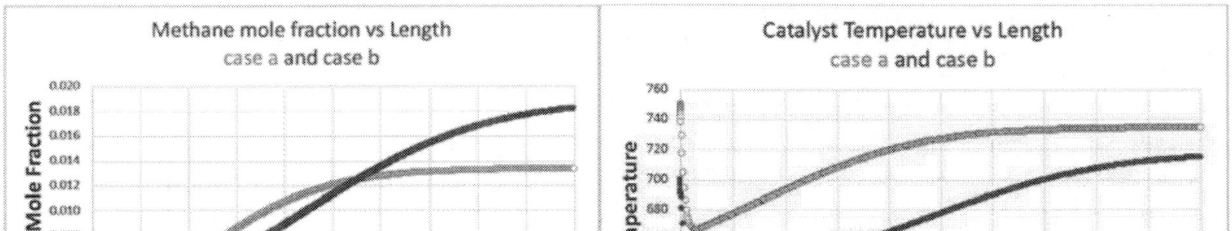

Figures 8A and 8B. With the selected microkinetic mechanism in a monolith configuration, DETCHEM CHANNEL predicts slightly higher methanation in the thinner "case a" catalyst (8A, left), which rases the temperature just enough to aid in the suppression of coke formation (8B, right).

For these cases, direct local temperature measurements are challenging, and the solution to promote a small amount of methanation is only found via simulation. Once again, accessible DETCHEM tools augment development by seeing what is happening inside the catalyst. The secret life of catalysts is no longer hidden, but now understood and problems overcome by changing geometry, operating conditions, thickness, or any other parameter available to the engineer.

Conclusion

The tradeoff of kinetics, pore diffusion, and heat transfer can lead to surprising, counterintuitive, and interesting findings. This secret life of catalysts might be promoting unwanted series or series-parallel reactions that are difficult to understand without supporting reaction engineering tools. In the case of the RWGS catalyst, a small amount of methanation (i.e., an unwanted side reaction) proved to be beneficial. Here, improving a catalyst to reduce methanation could in fact *exacerbate* carbon formation. For a practitioner, experimental observation alone would be challenging to understand why one system failed and the other did not. Development time and cost can be reduced by guiding efforts with useful analysis tools. The use of accessible reaction engineering tools (such as DETCHEM PBR and DETCHEM CHANNEL) speeds development and catalyst optimization, avoids pitfalls of experimentally missing the "sweet spot", reduces scaleup risk, and may solve vexing plant problems.

It is much easier to blame the catalyst for poor performance and move to the next interesting science experiment or accept the industrial expectation that performance deteriorates with scale up. Reaction kinetics are the Rosetta Stone to translate between catalyst geometries and reactor scales. Evaluating catalyst performance with tools that include realistic pore diffusion and reaction coupled with local heat transfer can reveal the true nature of a problem and provide clues on how to find practical solutions.

References

[1] Turing A. The chemical basis of morphogenesis. *Phil. Trans. R. Soc. London* 1952; **237**: 37–72. DOI: https://doi.org/10.1098/rstb.1952.0012

[2] Aris R. The Mathematical Theory of Diffusion and Reaction in Permeable Catalysts: The theory of the steady state. Oxford Clarendon Press; 1975.

[3] Fogler HS. *Elements of Chemical Reaction Engineering*. New Jersey Prentice Hal; 1986.

[4] omegadot software & consulting, GmbH. DETCHEM v3.0 2024. www.omegadot.software . Accessed March 6, 2024.

[5] HeatPath Solutions. https://heatpathsolutions.com . Accessed March 11, 2024.

[6] Delgado KH, Maier L, Tischer S, Zellner A, Stotz H, Deutschmann O. Surface reaction kinetics of steam- and CO_2-reforming as well as oxidation of methane over nickel-based catalysts. *Catalysts* 2015; 5(2): 871-904. DOI: https://doi.org/10.3390/catal5020871del

151d Ammonia As a Liquid Hydrogen Carrier: *Ab Initio* Multiscale Catalytic Synthesis/Cracking Reaction Modelling

Blaž Likozar, Andraz Pavlisic, Matic Pavlin, Taja Žibert, Aleksandra Zamljen, Sašo Gyergyek, Matej Huš

Introduction

Ammonia is gaining recognition as a carbon-free fuel for energy-intensive applications, particularly transportation, industry, and power generation. Due to its physical properties, high energy density of 3 kWh kg^{-1}, and high gravimetric hydrogen capacity of 17.6 wt%, ammonia is an efficient energy vector for green hydrogen, capable of mitigating hydrogen's storage, distribution, and infrastructure deployment limitations. Chemical storage in the form of ammonia provides an efficient and affordable solution for energy storage, which is currently a critical step in overcoming the intermittency of abundant renewable energy sources, with minimal or no environmental impact. By converting renewable energy into ammonia, it becomes possible to store energy long-term, as well as transport energy from areas with an excess of affordable renewable energy to where it is limited or expensive. With an already established reliable and flexible infrastructure for its distribution and an available feedstock (air and water), ammonia plays a crucial role in green transition towards sustainable energy future. However, the challenge for the global use of ammonia as an energy storage material remains more efficient, readily deployable production from abundant sources such as water and air in a decarbonized ammonia production process.

Catalysts play a vital role in ammonia synthesis by enabling the activation and reduction of an inert and stable nitrogen molecule. Since the use of an iron-based catalyst in the Haber-Bosch process, a significant progress has been made in the development of various catalysts for environmentally friendly ammonia synthesis. Molten iron catalysts, ruthenium-based catalysts and different metal nitride catalysts are common thermocatalysts, while noble metals (Au, Pt, Pd, Ru,...), non-noble metals (Fe, Mo-based catalysts) and non-metallic catalysts (MOF, ...) are used in sustainable electrochemical ammonia synthesis. Lastly, most promising photocatalysts are reported to be metal oxides, including titanium dioxide, bismuth oxyhalides, carbonaceous materials, metal sulfides, and bio-mimicking photocatalysts. In addition, different plasma catalysts showed a progress toward mild nitrogen reduction. On the other hand, the most promising catalysts for ammonia decomposition to obtain green hydrogen are different transition metals, such as Ru, Pd, Ni, Fe or Co, bimetallic catalysts, and catalysts that are suitable for both ammonia synthesis and decomposition.

Ammonia synthesis follows different mechanisms based on the process and catalyst used. In the dissociative pathway, dissociation of the nitrogen molecule is followed by successive hydrogenation steps on individual nitrogen atoms. In the associative alternating and distal mechanisms, the nitrogen molecule is adsorbed via the end-on configuration (vertical adsorption). In the distal pathway, the hydrogenation steps occur sequentially at the distal nitrogen atom, while in the alternating pathway they occur alternately at both nitrogen atoms until two ammonia molecules are formed. When used as an energy carrier, it is decomposed into a nitrogen molecule and a hydrogen molecule.

Methods in multiscale modelling

One of most common and efficient ways to study adsorbate-adsorbate and adsorbate-surface electronic interactions is using density functional theory. DFT modelling is useful in terms of catalyst design and optimisation. The subject of our research revolves around the synthesis and decomposition of ammonia (), a process that can be facilitated by a variety of catalysts. We chose to focus our efforts on modeling ruthenium (Ru) nanoparticles, employing a rigorous DFT analysis.

First, we began with a construction of a Ru slab model, followed by the execution of DFT calculations aimed at mapping out energy profiles and uncovering essential information about adsorption energies, preferred adsorption locations, and the complexity of the underlying reaction mechanisms. To carry out these calculations, we used GPAW, a computational tool that utilizes a descriptive approach involving plane waves (PW) with a specified energy cutoff of 450 eV and a grid of k-points set at 2×2×1. Additionally, we incorporated various correction techniques, including the dispersion correction DFT-D3 and dipole correction, to ensure the accuracy of our computational models. Moreover, we employed the Nudged Elastic Band (NEB) method, in conjunction with the Climbing Image approach, to pinpoint the transition states. This served as the foundation for our more extensive exploration into the potential role of Ru nanoparticles in ammonia synthesis.

Our research then took a significant step forward by expanding our focus to encompass Ru nanoparticles of different sizes and shapes, a comprehensive analysis that aimed to shed light on the influence of these factors on the synthesis and decomposition of ammonia.

Results

The purpose of the work was also to establish the connection between theoretical predictions and experimental work. DFT results were implemented into microkinetic models, which helped us to put DFT results into perspective. Using microkinetic models we were able to determine reaction conditions such as temperature and partial pressures as well as see the influence of number of active sites, calculate TOFs, selectivities, etc.

The results from DFT calculations (such as data on energetics and kinetics) were used in Kinetic Monte Carlo (kMC) simulations. In comparison with traditional microkinetic models which consider only average surface coverage, the kMC method also takes into account correlations, fluctuations, and the detailed spatial distributions of the adsorbates on the catalyst surface and their effect on the activation barriers and reaction rates. Catalytic activity of the catalyst in terms of turnover frequency, its surface composition and their dependency on wide variety of parameters (e.g. temperature, pressure, reactants composition etc.) is obtained as a result of the simulations.

Compared to all previously described methods and approaches, molecular dynamics (MD) simulations serve to study temporal evolution of the system by using Newtonian equations of motions. By applying principles of classical mechanics and statistical thermodynamics, MD simulations provides insights into various properties, such as temperature, pressure, and structural changes, allowing us to understand the dynamic behaviour of materials at the atomic level. MD simulations are robust and computationally cheaper than DFT calculations, enabling study of larger systems over extended time periods.

One of the promising class of ammonia sorbents are metal halides supported on materials such as silica and different zeolites. In this study, we conducted classical MD simulations using existing reactive force fields, with the exception of $MgCl_2$ parameters, which we developed ourselves. Our study aimed to: (i) construct amorphous silica structures with varying pore sizes, (ii) investigate the permeation of $MgCl_2$ into these pores based on different pore dimensions, (iii) assess the stability of $MgCl_2$ supported on amorphous silica under high temperatures and pressures, and (iv) explore the mechanisms of ammonia absorption on $MgCl_2$ supported on amorphous silica at high temperatures and pressures.

Despite technological maturity, understanding the reaction mechanism and the interplay of the surface properties is still an ongoing topic. Thus, the exact mechanism of the reaction pathway and the interplay of the catalytic surface have received much attention in academia in recent years. While multi-scale hierarchical approach was already successfully applied to a compelling industrial process, where catalytic heterogenous system is captured from micro to macro level, it is hard to believe that this will become a main stream of solving industrial processes due to high computational cost of the methods.

On the other hand, correlation equations (CE) have a long history of use in chemical engineering. They are collection of dimensionless numbers and coefficients that gives fast and meaningful prediction of mass and heat transfer. The advantage of CE is that they project three dimensional phenomena to one dimension by averaging physiochemical properties which can be used in much simpler continuum fluid dynamic models such as plug flow reactor (PFR). However, they have to be experimentally determined for each specific system which in case of industrial scale reactors is prohibitively expensive. Moreover, CE validity is strictly bonded to the specific range of the operational conditions of the system. On the other hand, all the local phenomena which can be used to calculate CE can be obtained with aid of CFD. Thus, mass and heat transfer calculated with CFD simulation are linked to pseudo-homogeneous and heterogeneous continuum models. With this technique we can create modules for fast and targeted optimization of industrial processes.

Future outlook

Photocatalysis is a greener and sustainable alternative to the conventional Haber-Bosch process, as it enables the nitrogen reduction reaction (NRR) under mild conditions. Photocatalytic first-principles nitrogen reduction reaction is performed using different approaches, including delta SCF, time-dependent density functional theory and maximum overlap method. In our study, we investigated the nitrogen reduction reaction over the Ru-doped TiO_2 clusters using the maximum overlap method to consider exited states. We obtained the most promising nitrogen reduction mechanism in the lowest excited state and compare the results to the ground state mechanism. The determined adsorption modes of initial (N_2, H, 2H) and final (NH_3) precursors of the NRR showed that end-on adsorption configuration of nitrogen is the most promising, therefore we focused on evaluating the dissociative, associative distal and alternating pathways. An associative enzymatic mechanism was due to unfavourable side-on adsorption of nitrogen excluded from further studies. The photocatalytic NRR determined by first-principles studies is a good starting point for building a multiscale model of photocatalytic ammonia synthesis, which contributes to the understanding of sustainable ammonia synthesis and decomposition.

195a Producing Alternative Feedstocks through Shell Fiber Conversion Technology

Sheila Lucas

Shell Fiber Conversion Technology is a bolt-on technology that converts lignocellulosic biomass, such as wet distillers grains (WDG) from corn ethanol plants, into enhanced-protein animal feeds, distillers corn oil (DCO) and cellulosic ethanol. Distillers corn oil and ethanol will be important alternative feedstocks helping the energy transition. The process uses a reusable acid, and the low-temperature operation is intrinsically non-fouling. The technology has been designed to achieve benchmark levels of operability, reliability, and energy efficiency.

195b Sustainable Manufacturing of Acrylonitrile

Anthony Martinez

Trillium Renewable Chemicals is scaling a two-step thermochemical process that converts a biobased feedstock (glycerol) into acrylonitrile and valuable by-products, acetonitrile and HCN. Acrylonitrile (ACN) is used to manufacture a wide-range of materials (e.g., acrylonitrile-butadiene-styrene (ABS) plastic, carbon fiber, acrylic fiber, nitrile butadiene rubber (NBR), acrylamide, specialty amines) with a wide range of end uses including aerospace, automotive, apparel, toys, energy, personal care products, pharmaceuticals, consumer electronics, and wastewater treatment. The presentation will focus on the progress toward the ultimate goal: de-risking Trillium's bio-ACN™ process for commercialization while achieving stretch targets for sustainability through a combination of low-carbon raw materials and innovative process intensification concepts as well as meeting unit cost economics at parity or below the incumbent process to produce acrylonitrile.

195c Modular Pilot/Demonstration Plants: A Cornerstone for Accelerating the Global Transition to Clean Energy

Thomas Schafer

In the current climate of escalating environmental concerns and an imperative to mitigate climate change, this paper underscores the pivotal role of modular pilot and demonstration scale plants as a linchpin in expediting the global transition to clean energy solutions.

Modular pilot and demonstration scale plants offer a unique and transformative approach to energy research and development. Their capacity to optimize innovation processes, reduce project timelines, and manage inherent risks makes them an indispensable asset in the clean energy landscape.

Beyond their streamlined approach to research and development, these systems enable comprehensive sustainable feedstock experimentation, tailored to the requirements of each project. Moreover, modular pilot plants play a vital role in the validation and scale-up of clean energy processes, thereby enhancing efficiency, reducing costs, and accelerating the deployment of clean energy solutions.

In this paper, we will discuss real-world success stories showcasing the use of modular pilot and demonstration scale plants in biomass-to-chemicals, regenerative agriculture-to-food-and-biofuel, biomass-to-biofuel, and carbon capture systems.

By attending this presentation, you will leave with a better understanding of when you should consider incorporating a modular pilot scale plant within your overall project development process, important design and project delivery considerations, and associated commercialization benefits.

152a Decarbonizing ASEAN By 2050: From the Lens of a Hydrogen Economy

Archana Kumaraswamy, Sushant Garud, Iftekar Karimi, Farooq Shamsuzzaman

Decarbonizing ASEAN is a challenge due to its heavy reliance on fossil fuels and rapidly growing energy demand. Hydrogen is widely believed to be a promising vector for decarbonization. Understanding the evolution of a hydrogen economy is a complex optimization problem that depends on a variety of factors due to following reasons – (1) temporal variation in cost of H_2 production technologies, (2) varying emissions reduction capabilities of different H_2 production technologies, (3) process efficiencies, (4) resource availabilities [1-2] (5) projected renewable electricity penetration, (6) hydrogen transportation costs, and (7) spatial variation in capital costs. This study uses an in-house tool (Hydrogen Economy Assessment & Resource Tool, HEART) [3], developed in an earlier study and advanced further in the present study, to find the minimal cost hydrogen roadmap for ASEAN to achieve net zero emissions by 2050. A decarbonization profile is assumed for this purpose. Based on the assumed declining renewable electricity cost profiles expected from technological advances, this hydrogen roadmap determines the optimal hydrogen production technology mix (under various resource constraint scenarios), production locations, and supply-chain network. It has been shown that ASEAN's renewable energy potential is sufficient to fully meet the total power demand of the member countries if a regional power grid is put in place and active energy storage and management is employed. The region has surplus renewable electricity potential to produce green hydrogen even after meeting the total power demand. With realistic limits on the availability of biomass, it is not possible to meet the carbon emission reduction targets with hydrogen produced only from natural gas and biomass (with carbon capture and sequestration) beyond 2045, even if natural gas is unlimited. It is found that reduction in upstream supply chain emissions with time will allow more time for green hydrogen to attain cost parity with blue hydrogen. Green hydrogen can be cost effective compared to blue hydrogen produced from LNG in the early years when the hydrogen demand is low, as it is not impacted by economies of scale. Using hydrogen FCVs to decarbonize the transport sector will inflate the average cost of hydrogen, due to costs associated with compression of hydrogen to 700 bar. Prioritizing energy security and sustainability has minimal incremental costs compared to importing LNG for blue hydrogen production. Optimal green hydrogen production locations in ASEAN include Myanmar and Cambodia, as they have the lowest renewables levelized cost of electricity (LCOE) in the region. Malaysia and Indonesia, although endowed with solar energy, use biomass resources or LNG for hydrogen production instead. This is due to relatively higher solar LCOE. It is found to be cost optimal for Singapore to import hydrogen rather than produce hydrogen using imported LNG. ASEAN's hydrogen economy is expected to evolve from centralized in the early years to one that is decentralized in the later years, due to the large share of production cost of the total landed cost in the early years. Ammonia as a hydrogen carrier is found to be cost-optimal for Myanmar because it is located more than 2000 km away from most demand sites. In summary, this study develops a holistic and comprehensive hydrogen roadmap to decarbonize ASEAN by 2050.

[1] A. Vidinopoulos, J. Whale, and U. Fuentes Hutfilter, "Assessing the technical potential of ASEAN countries to achieve 100% renewable energy supply," *Sustainable Energy Technologies and Assessments*, vol. 42, p. 100878, Dec. 2020, doi: 10.1016/j.seta.2020.100878.

[2] T. Ahmed, S. Mekhilef, R. Shah, and N. Mithulananthan, "Investigation into transmission options for cross-border power trading in ASEAN power grid," *Energy Policy*, vol. 108, pp. 91–101, Sep. 2017, doi: 10.1016/j.enpol.2017.05.020.

[3] X. Hong *et al.*, "Hydrogen Economy Assessment & Resource Tool (HEART): A python-based tool for ASEAN H2 roadmap study," *International Journal of Hydrogen Energy*, 2022, doi: 10.1016/j.ijhydene.2022.05.036.

152b Horses for Courses: Matching Modeling Techniques for Varying Aspects of Energy Transition Solutions

Ravindra Aglave, Shivdeep Gaagat, Barbara Neuhierl

Energy transition is an urgent but delicate topic requiring balancing of many engineering, business and social factors, at times some of which are contradicting. The application of the Horses modeling method is certainly a tool in the toolchest of solutions. Not a particular single modeling method can provide all the answers, nor a single solution strand of transition is sufficient. In this talk, we would like to present an ecosystem of modeling methods for an ecosystem of energy transition options. All of these are housed under the DEGREE framework of sustainability at Siemens. Examples of use cases include H2 transport, fuel cells, H2 storage, and charging stations. Methods of modeling include molecular models, process models, system models, and life cycle management solutions.

Reference

1. Simulation for energy transition solutions

SPRING24
+20TH GCPS
A Joint AIChE and CCPS Meeting

Quantifying Sustainability Impacts of Post-Combustion Carbon Capture Using Process Simulation

Clinton M. Duewall
Texas A&M University; Bryan Research & Engineering, LLC
Artie McFerrin Department of Chemical Engineering, College Station, TX, USA
clint.duewall@bre.com

Mahmoud M. El-Halwagi
Texas A&M University; Texas A&M Engineering Experiment Station Gas and Fuels Research Center
Artie McFerrin Department of Chemical Engineering, College Station, TX, USA
el-halwagi@tamu.edu

Prepared for Presentation at
American Institute of Chemical Engineers
2024 Spring Meeting and 20th Global Congress on Process Safety
New Orleans, LA
March 24-28, 2024

Quantifying Sustainability Impacts of Post-Combustion Carbon Capture Using Process Simulation

Clinton M. Duewall
Texas A&M University; Bryan Research & Engineering, LLC
Artie McFerrin Department of Chemical Engineering, College Station, TX, USA
clint.duewall@bre.com

Mahmoud M. El-Halwagi
Texas A&M University; Texas A&M Engineering Experiment Station Gas and Fuels Research Center

Keywords: Computing and Systems Engineering, Sustainability & Environment

Abstract

The International Energy Agency sees carbon capture, utilization, and storage (CCUS) as a key route for decarbonization in the energy transition. Researchers have considered many different CCUS technologies, and each application to an existing facility has a different effect on the process. Changes in each facet of sustainability (economic, environmental, and social) should be measured to determine which projects are the most sustainable developments. With hundreds of CCUS projects in development but few operational, computer modeling tools like process simulation remain one of the most effective ways of predicting the overall impact of carbon capture on a given process. This work uses process simulation to quantify the impact of amine-based carbon capture on the sustainability of a combined-cycle natural gas power plant. Economic analysis includes capital and operating cost estimates from simulated parameters. Life cycle analysis and model predicted emissions quantify environmental and societal impacts. The results justify recommendations to improve the sustainability of existing technology and identify better targets for the application of carbon capture.

1 Introduction

Sustainable designs balance economic, environmental, and social criteria to provide value to the shareholder, the planet, and the population. Industry has begun to replace purely economic metrics for success for sustainability, leading engineers to study methods to compare different designs using this new metric.

Many studies have analyzed the economics of monoethanolamine (MEA) carbon capture for both coal[1] and NGCC[2] power plants. Most of these studies only consider economic or environmental criteria, comparing capital and operating costs with waste and emissions.[3] Several studies perform exhaustive life cycle analyses for environmental impact without considering whether the design in question is economically viable.[4] The purpose of this study is to quantify the sustainability impact of carbon capture through process simulation, a tool with which many process engineers have access and familiarity. Sustainability metrics are used to augment traditional economic indicators for different potential retrofit designs for an example 100 megawatt NGCC power plant.

1.1 Sustainability Metrics

Sustainability metrics are methods of measuring progress towards one or more of the three dimensions of sustainability.[5] These metrics are typically classified by "dimensionality", where a one-dimensional (1D) metric affects a single aspect of sustainability, either economic, environmental, or social.[6] Three-dimensional (3D) metrics are generally considered the most appropriate unless attributes of specific processes make a lower dimensional metric important.

One example of a 3D metric applicable to most industrial processes is material intensity, shown in Equation 1. Material intensity directly compares the use of any process material input with plant output. For a power plant, this could represent the volume of natural gas used to produce a kilowatt-hour of electricity.

$$MI = \frac{\text{Mass Material Used}}{\text{Output}} \tag{1}$$

Choosing the right set of sustainability metrics is essential to their use. This requires knowledge of the process and company objectives. Previous research has differing conclusions on how to make decisions using metrics. Some works suggest aggregating all applicable metrics together into a single measure of sustainability,[7,8] while others recommend using a small set of independent metrics separately. These metrics should collectively cover all three dimensions of sustainability, ideally as 3D metrics themselves, and adequately represent the process being studied.[6]

1.2 NGCC Carbon Capture

As demands for both energy and reduced carbon emissions grow, industry is turning to an old technology to be used in a relatively new way. The new challenge, however, is designing CO_2 removal units to handle immense volumes of gas at elevated temperatures near atmospheric pressure. Partial pressure is the main driving force for CO_2 pickup in amines. Flue gas from a natural gas combined cycle (NGCC) power plant has a CO_2 partial pressure of about 0.04 atm. This driving force is orders of magnitude lower than the typical natural gas treatment application.

As a result of these design challenges, flue gas carbon capture is expensive from both a capital and operating perspective. The general process flow diagram, shown in Figure 1, is much like traditional natural gas treating with amines. CO_2 rich flue gas is contacted with amine solvent in the absorber. The resulting CO_2 rich amine is pumped to the regenerator where a steam driven reboiler strips the CO_2 from the solvent, preparing it to reenter the absorber.

Carbon capture plants differ from typical natural gas treating units since the absorber is typically lower pressure than the regenerator. Running the absorber near atmospheric pressure leads to challenges with solvent overhead losses. These issues are most severe for MEA but are still significant with lower vapor pressure amines like piperazine and aminomethyl propanol (AMP). Regenerator pressures are typically the same between gas treatment facilities and carbon capture units. Lowering the pressure further in the regenerator typically results in more compression costs than reboiler duty savings.

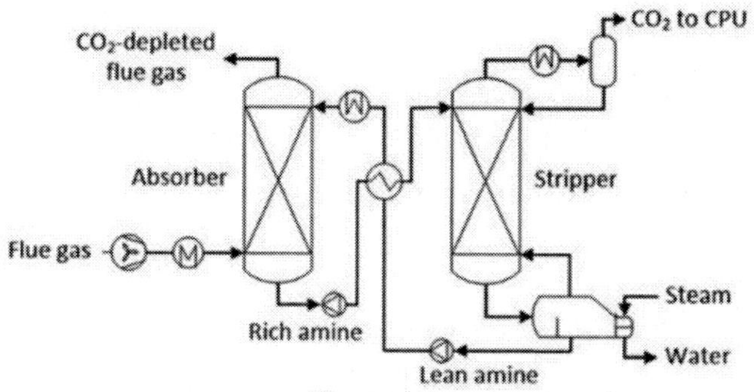

Figure 1: Amine Unit Block Flow Diagram

NGCC plants are some of the most flexible sources of power generation in the grid mix and are frequently required to modulate power supply to meet consumer demand.[9] Design and operations changes to handle changing power demand have been studied, but mostly for the purpose of mitigating operational upset conditions, not observing sustainability effects.[10]

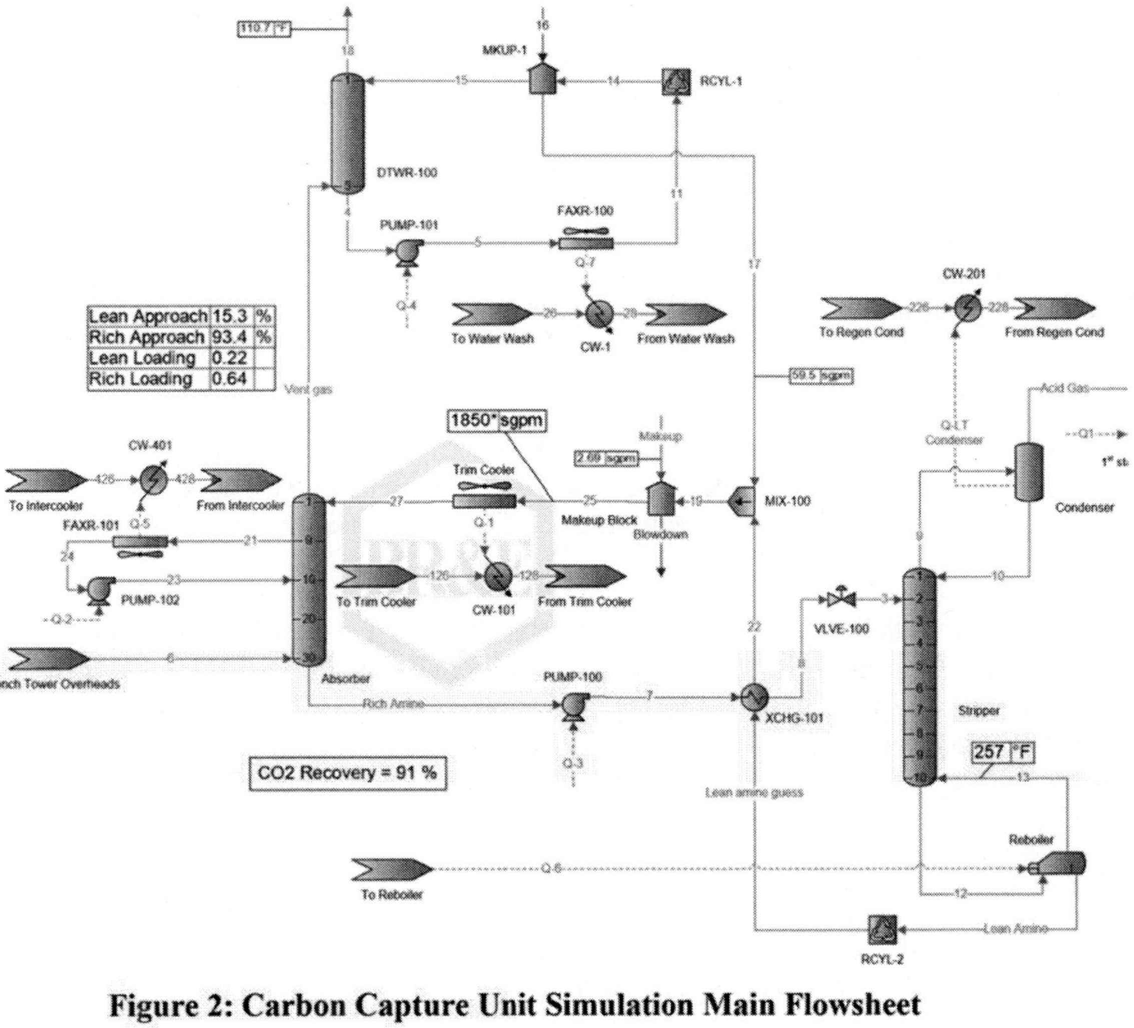

Figure 2: Carbon Capture Unit Simulation Main Flowsheet

2 Methodology

The following section details the steps used to generate the results of this work. General characteristics of the process models are explained, along with details on each sustainability metric used. Finally, key assumptions made in the economic analysis are listed with assumed raw material and product pricing.

2.1 Process Modeling

ProMax[11] is used to model the NGCC power generation unit, carbon capture unit, and utilities. The main flowsheet for the carbon capture unit is depicted in Figure 2. Modeling the power plant facilitated heat integration between the plant and the capture unit. Low pressure steam from the heat recovery section of the power plant is used to drive the solvent regenerator. Steam returns from the reboiler as a saturated liquid, decreasing cooling water demand for the NGCC unit.

The carbon capture unit is modeled as an absorber with separate water wash sections. These are an equivalent model to a single vessel with multiple packed sections. The CESAR-1 solvent blend of 27 wt% AMP and 13 wt% piperazine is used. The absorber is packed with 60 feet of Sulzer Mellacarbon 250 Y structured packing. Each water wash section is an additional 30 feet of structured packing. The absorber and water wash sections each use ProMax's TSWEET column model to accurately predict the rate of CO_2 absorption. Pumps and compressors are assumed to be driven by electric motors supplied by power from the NGCC unit.

2.2 Experimental Designs

Four potential amine unit designs are considered along with a base case of the NGCC unit without carbon capture. Each unit uses a different absorber configuration with the same regenerator and auxiliary equipment and achieves 90% recovery of the CO_2 fed. Case 1 uses a 30-foot diameter amine absorber designed to accept the full flue gas stream from the NGCC unit running at 100 MW with a single water wash as shown in Figure 2.

Case 2 is identical to Case 1 but adds a second identical water wash section to the absorber that is purged from the unit instead of being recycled into the solvent loop. Cases 3 and 4 are identical to Cases 1 and 2, but the absorber is reduced to 22 feet to accept 60% of the flue gas stream from the NGCC unit running at 100 MW.

An example utilization curve is used to test each unit running off peak power production to simulate variable grid power demands. The curve is designed to average to a capacity factor of 60%, matching the average utilization of NGCC plants in the United States. It contains periods of shutdown, maximal power production, and several different turndown ratios over a week of operation. The utilization curve is shown in Figure 3 in the Results and Analysis section of this paper.

2.3 Sustainability Analysis

The ultimate measure of design merit in this study is the sustainability-weighted return on investment metric (SWROIM).[12] This metric combines a requirement for economic feasibility while allowing a process specific sustainability metrics to influence which solution is selected. Calculating SWROIM for each potential design requires a value for each sustainability plus return on investment (ROI). ROI is defined as annual net after-tax profit divided by the total capital cost. Each metric was normalized on a scale of zero to one, where zero is the worst performing design alternative and one is the best.

$$\text{SWROIM} = \text{ROI}\left[1 + \sum_{i=1}^{N_{\text{Metrics}}} w_i \left(\frac{\text{Metric}_i}{\text{Metric}_i^{\text{Target}}}\right)\right] \quad (2)$$

The set of metrics includes total power production weighted 0.1, CO_2 emissions weighted 0.25, and material intensity for water, amine, and natural gas, weighted at 0.1 each. Chosen metrics favor environmental and social dimensions of sustainability, since economics are well represented with the ROI term in SWROIM. Net power production, Equation 3, measures the amount of electrical power that can be produced with the carbon capture unit running. This metric serves as a ceiling on a major revenue stream and has significant social implications, since it effectively reduces a population's power supply while the carbon capture unit is running. Material intensities are calculated with Equation 1, using specific material consumption per unit of energy production. CO_2 emissions, for this study, are concentrated to just the vent gas carbon dioxide emissions.

$$\text{Net Power} = P_{\text{Gas Turbine}} + P_{\text{Steam Turbines}} - W_{\text{Pumps}} - W_{\text{Compressors}} \quad (3)$$

ROI requires an estimate of total capital investment, revenue, and operating cost. This study uses the six-tenths rule[13] to scale the AMP-piperazine capture unit capacity to the nameplate power plant size from Manzolini et al.,[14] CEPCI economic indicators,[15] and currency exchange rates are used to adjust the capital cost to October 2023 USD. The equipment cost of a single-wash column is assumed to be 70% of the cost of the dual-wash column. Depreciation was calculated using the straight-line method with no salvage value over a period of 10 years.

The NGCC carbon capture facility is assumed to be located on the Texas Gulf Coast, providing an opportunity for CO_2 sales for enhanced oil recovery. A ton (2000 lbs.) of CO_2 for enhanced oil recovery typically sells for ~40% of the oil price per barrel.[16] Assuming a West Texas Intermediate price of $70/bbl, any CO_2 captured and compressed can be sold for $25.40/tonne. The US Internal Revenue Code Section 45Q creates a tax credit of up to $60/tonne for the first 12 years a facility is in service depending on a long list of contingencies that must be followed. Net electrical power production is sold at $0.08/kWh.

Feed natural gas and fixed operating and maintenance costs dominate the operating expenditure of the plant. AMP and piperazine are assumed to cost $8.88/kg and $6.66/kg, respectively. Fresh process water is available at $2.68 per thousand gallons, wastewater disposal costs $1.65 per thousand gallons, and cooling water costs $0.27 per thousand gallons. Natural gas is available at $3/MMBtu, and the corporate income tax is assumed to be 21%.

3 Results and Analysis

Under the economic assumptions made and a capacity factor of 100%, each design produced a favorable ROI greater than 17%. Full capacity carbon capture with a single water wash performed the best by both ROI (24.3%) and SWROIM (30.7%). While operating costs are significantly higher, the additional revenue stream of CO_2 sales, depreciation, and federal tax credits increases the plant's annual net after-tax profit from $41 million to $48-56 million. Data to corroborate these observations is available in Table 1. Here total equipment cost (TEC), total capital investment (TCI), annual gross profit (AGP), annual net after-tax profit (ANP), and ANP per tonne of CO_2 captured are compared for each design and each capacity factor.

Over the 10-year operating period, the increased capital cost of a second water wash section outweighed the savings on solvent retained. This design was originally expected to perform better by SWROIM. Based upon the sustainability metrics chosen, the increased water intensity superseded the decrease in amine intensity.

Table 1: Design Economic Summary

		Units	Base	60% CC 1WW	Full CC 2WW	60% CC 2WW	Full CC 1 WW
Fixed Costs	TEC	$MM	0	18.02	25.92	19.08	24.49
	TCI	$MM	0	43.31	62.29	45.85	58.84
	Depreciation	$MM/yr	0	4.33	6.23	4.58	5.88
100% Capacity Factor	AGP	$MM/yr	51.87	47.55	47.07	47.81	46.60
	Tax Credits	$MM/yr	0	10.30	17.25	10.30	17.25
	ANP	$MM/yr	40.98	48.77	55.74	49.03	55.30
	ROI	%	N/A	18.0%	23.7%	17.5%	24.3%
	Specific Net ANP	$/tonne	0	45.37	51.34	46.88	49.82
60% Capacity Factor	AGP	$MM/yr	31.37	30.59	30.15	30.75	29.87
	Tax Credits	$MM/yr	0	8.37	8.70	8.37	8.70
	ANP	$MM/yr	24.78	33.44	33.83	33.62	33.54
	ROI	%	N/A	20.0%	14.5%	19.3%	14.9%
	Specific Net ANP	$/tonne	0	62.11	62.40	63.39	60.37

The results thus far do not differ from available literature. When a dynamic demand with a capacity factor of 60% is used, however, the most economic and sustainable design completely reverses. At the lower capacity factor, the difference between the amount of CO_2 captured narrows, so the difference in revenue from CO_2 sales, and the Section 45Q tax credit narrows as well. Here, lower capital investment of the smaller capture unit raises the ROI and SWROIM above that of the full capture unit. Just as in the 100% capacity factor case, single water-wash designs outperformed the dual-wash designs. The 60% capacity factor single water-wash design produced the best performance with an ROI of 20.0% and SWROIM of 26.4%.

Figure 3 shows how variations in power demand common to NGCC units affect carbon capture unit performance. The example utilization case on 2/2 shows a period where the carbon capture unit must be shut down to meet power demand. When 90 MW is required on 1/31 and 2/3, the full capture unit must shut down to deliver the required power, while the 60% capture unit is still running. For this example utilization curve, the overall difference in CO_2 recovery is only 3%.

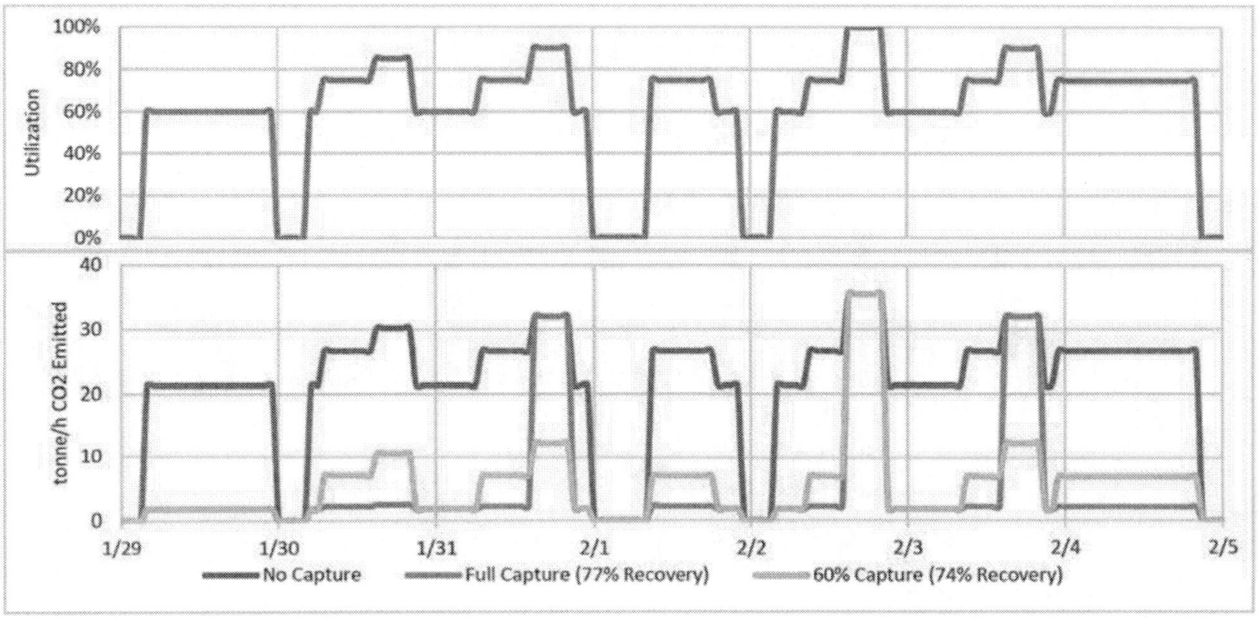

Figure 3: Variable Demand CO_2 Emissions

Another key takeaway from these results is each design lowered earnings before taxes, depreciation, and amortization (EBITDA) by $4-5 million. This is also observed in Table 1 where the AGP is lower for each design case than the base NGCC plant without carbon capture. To investigate this further, several dollar-amounts for the Section 45Q tax credit are tested and their results in Figure 4 and Figure 5 display the sensitivity of ROI and SWROIM on the tax credit. Unsurprisingly both ROI and SWROIM are linearly dependent upon the magnitude of the Section 45Q tax credit.

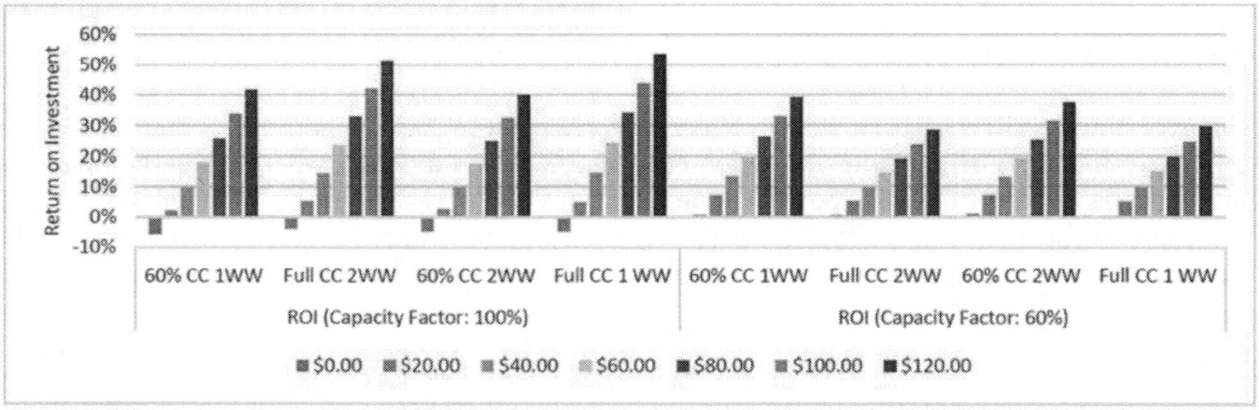

Figure 4: Carbon Tax Credit Sensitivity on ROI

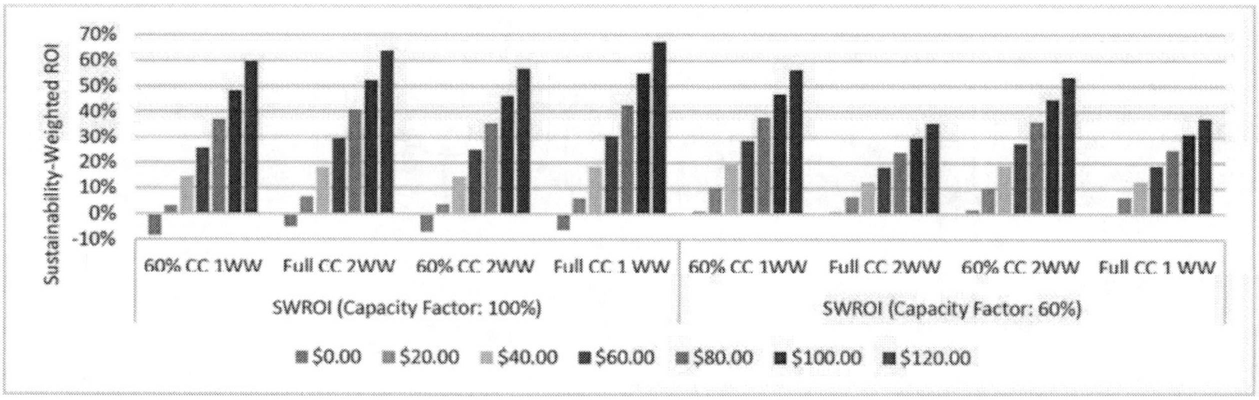

Figure 5: Carbon Tax Credit Sensitivity on SWROIM

Full capture designs are more sensitive to the Section 45Q tax credit than 60% capture designs for capacity factor of 100%. When the capacity factor is reduced to a more realistic 60%, however, the opposite trend is observed. Regardless of the design choice, the potential for future legislation to supersede Section 45Q will add a level of risk to any carbon capture retrofit project.

Since the other sources of costs and revenue are mostly tied to commodity markets, the potential variability in revenue or cost due to natural gas prices, CO_2 sale price, or solvent price are unlikely to cause as much effect. Eliminating the sale of CO_2 for enhanced oil recovery, the ROI only drops 7% for each design, resulting in ROIs of 10-15% and SWROIMs of 12-19%.

4 Conclusions

The results of this study show that the relative sustainability of post combustion carbon capture of a natural gas combined cycle power plant designs can be quantitatively compared using process simulation. The methodology used in this work can be applied to different point-source carbon emissions to determine which technology is most sustainable. This study also provides an example of important sustainability metrics for flue gas carbon capture.

The case study showed that different power plant utilization rates alter which design is most sustainable. The design built to handle 60% of nameplate flue gas outperformed the full capture unit when the plant saw a variable demand averaging to a capacity factor of 60%, while the opposite was true for operation at base power production. The economics, and therefore sustainability, of carbon capture are inexorably tied to tax credit incentives, with each design producing a negative return on investment, and sustainability-weighted return on investment metric.

Based on the weights and metrics chosen, the increased water demand and capital cost of adding a second water wash to the amine contactor reduced the sustainability of the design. This result would likely reverse if either more weight is applied to amine pollutant emissions, or the plant lifetime is increased beyond 10 years.

One complicating factor to note is the Section 45Q tax credits that are responsible for overcoming the additional operating cost may only be taken for 12 years. After this point, companies will lose money each time they run their carbon capture unit. Barring an exorbitant rise in CO_2 demand, operators will likely be forced to shut down these units to maintain economic viability in a commodity market.

5 References

[1] Koiwanit J, Piewkhaow L, Zhou Q, et al. A life cycle assessment study of a Canadian post-combustion carbon dioxide capture process system. *Int J Life Cycle Assess*. 2014;19(2):357-369. doi:10.1007/s11367-013-0649-2

[2] Mohammed RK, Farzaneh H. Life Cycle Environmental Impacts Assessment of Post-Combustion Carbon Capture for Natural Gas Combined Cycle Power Plant in Iraq, Considering Grassroots and Retrofit Design. *Energies*. 2023;16(3):1545. doi:10.3390/en16031545

[3] Rao AB, Rubin ES. A Technical, Economic, and Environmental Assessment of Amine-Based CO $_2$ Capture Technology for Power Plant Greenhouse Gas Control. *Environ Sci Technol*. 2002;36(20):4467-4475. doi:10.1021/es0158861

[4] Spath PL, Mann MK. *Life Cycle Assessment of a Natural Gas Combined Cycle Power Generation System*.; 2000:NREL/TP-570-27715, 776930. doi:10.2172/776930

[5] Sikdar SK, Mukherjee R, Sengupta D. *Measuring Progress Towards Sustainability: A Treatise for Engineers*. 1st ed. 2017. Springer International Publishing : Imprint: Springer; 2017. doi:10.1007/978-3-319-42719-5

[6] Martins AA, Mata TM, Costa CAV, Sikdar SK. Framework for Sustainability Metrics. *Ind Eng Chem Res*. 2007;46(10):2962-2973. doi:10.1021/ie060692l

[7] Sikdar SK. On aggregating multiple indicators into a single metric for sustainability. *Clean Technol Environ Policy*. 2009;11(2):157-161. doi:10.1007/s10098-009-0225-4

[8] Sikdar SK, Sengupta D, Harten P. More on aggregating multiple indicators into a single index for sustainability analyses. *Clean Technol Environ Policy*. 2012;14(5):765-773. doi:10.1007/s10098-012-0520-3

[9] Diego ME, Akram M, Bellas J, Finney KN, Pourkashanian M. Making gas-CCS a commercial reality: The challenges of scaling up. *Greenh Gases Sci Technol*. 2017;7(5):778-801. doi:10.1002/ghg.1695

[10] Domenichini R, Mancuso L, Ferrari N, Davison J. Operating Flexibility of Power Plants with Carbon Capture and Storage (CCS). *Energy Procedia*. 2013;37:2727-2737. doi:10.1016/j.egypro.2013.06.157

[11] ProMax 6.0. Bryan Research & Engineering LLC. Bryan, TX, USA. 2024.

[12] El-Halwagi MM. A return on investment metric for incorporating sustainability in process integration and improvement projects. *Clean Technol Environ Policy*. 2017;19(2):611-617. doi:10.1007/s10098-016-1280-2

[13] El-Halwagi MM. *Sustainable Design through Process Integration: Fundamentals and Applications to Industrial Pollution Prevention, Resource Conservation, and Profitability Enhancement*. Second edition. Elsevier; 2017.

[14] Manzolini G, Sanchez Fernandez E, Rezvani S, Macchi E, Goetheer ELV, Vlugt TJH. Economic assessment of novel amine based CO2 capture technologies integrated in power plants based on European Benchmarking Task Force methodology. *Appl Energy*. 2015;138:546-558. doi:10.1016/j.apenergy.2014.04.066

[15] Anonymous. Economic Indicators. *Chem Eng*. 2024;131(2):48.

[16] Edwards RWJ, Celia MA. Infrastructure to enable deployment of carbon capture, utilization, and storage in the United States. *Proc Natl Acad Sci*. 2018;115(38). doi:10.1073/pnas.1806504115

Simulation-Based Approach for Assessing Risk Factors in Offshore Wind-Integrated Hydrogen Production Processes

Md Tanjin Amin, Mahmoud El-Halwagi, Faisal Khan, Ahmed Elkady

Green hydrogen production via water splitting uses fresh water and renewable electricity to create hydrogen while emitting oxygen as its sole byproduct. In offshore locations, seawater is readily available but requires desalination and deionization for traditional electrolysis. These facilities can also tap into offshore wind farms to generate hydrogen for electricity storage during periods of high wind yield, storing it for future use or converting it back into electricity using fuel cells. In the current study, we have employed a simulation-based approach to delve deeper and further analyze and assess the operational risk factors. First, we utilized Aspen Custom Modeling to custom-create models that outline the equations for the three key technologies involved: electrolysis, wind turbines, and desalination. These technologies were, then, interconnected using Aspen Plus, along with essential components such as pumps, gas compressors, heat exchangers, and pressure exchangers. We examined how operating conditions, including power, current density, and pressure, affected electrolysis efficiency. Additionally, we evaluated alterations in the seawater feed rate, desalination, and deionization reject rates under varying operational conditions, examining their effects on membrane thinning and, consequently, gas crossover. Our study culminated in the identification of the optimal operating range for specific membrane thickness, with due consideration for both safety and electrolysis efficiency.

114a Potential of Solid-State Anaerobic Digestion for Biogas Production from Poultry Waste

Faryal Fatima, Raghava Kommalapati

Conventional liquid anaerobic digestion typically operates with a total solids content below 15%, whereas solid-state anaerobic digestion involves a total solids content exceeding 15%. In comparison to liquid anaerobic digestion, solid-state anaerobic digestion offers several advantages, including reduced energy requirements for heating and the production of a digestate with lower moisture content. The primary objective of this project was to evaluate the solid-state anaerobic digestion of poultry waste, encompassing organs, intestines, and offal meat for biogas production. The solid-state anaerobic digestion of poultry waste was conducted using biochemical methane potential (BMP) assays. The poultry waste and the inoculum were characterized by measuring parameters such as total solids, volatile solids, and moisture content. For precise analysis of the poultry waste, the Thermo Scientific™ FLASH 2000 CHNS/O Analyzer was used to determine the carbon, nitrogen, hydrogen, oxygen and sulfur content. The BMP assays were performed in 250 mL bottles with a working volume of 100 mL. The substrate-to-inoculum ratio was set at 0.5. Biogas production was measured using the liquid displacement method, with sodium hydroxide serving as a buffering agent during the experiments. The results of the study demonstrated that solid-state anaerobic digestion of poultry waste yielded 760 mL of biogas. This finding underscores the potential benefits and efficiency of using solid-state anaerobic digestion for biogas production from poultry waste. Valorizing poultry waste through solid-state anaerobic digestion not only contributes to reducing environmental burdens but also provides an opportunity to harness a valuable resource for energy generation. This research promotes a more sustainable approach to waste management and explores the practical application of biogas production from poultry waste through solid-state anaerobic digestion.

114b Utilizing a Modular Induction Reactor, on-Demand Ammonia Synthesis Harmonizes Seamlessly with Intermittent Renewable Energy Patterns.

Janvit Terzan, Anja Sedminek, Žiga Ponikvar, Matej Huš, Darko Makovec, Sašo Gyergyek, Blaž Likozar

Ammonia (NH_3) is distinguished as an ideal hydrogen carrier by its high volumetric hydrogen energy density of 10.7 kg H_2/100 l. This property, coupled with its ease of liquefaction, handling, storage and transport, makes ammonia a promising component of a sustainable energy ecosystem. Despite these advantages, about 90 % of the world's ammonia production still relies on the centuries-old Haber-Bosch process, a thermocatalytic conversion of nitrogen (N_2) and hydrogen (H_2) at high temperatures (above 300 °C) and high pressure (above 15 MPa). The Haber-Bosch process is one of the largest industrial chemical processes in the world and has been hailed as a groundbreaking invention of the 20th century, winning three Nobel Prizes in Chemistry.

In our quest to modernise and optimise ammonia synthesis, we present a breakthrough approach focused on on-demand, delocalised ammonia synthesis. This innovative process exploits the remarkable hydrogen storage capacity and favourable physical properties of ammonia to address the challenges associated with renewable energy integration, hydrogen storage and efficient energy conversion.

The breakthrough in our approach is the operation of the reactor where the reactants and products are kept at near room temperature, only 3 cm from the catalyst bed. This proximity not only contributes to safety and efficiency, but also minimises energy losses throughout the synthesis process. By operating the reactor under low hydrogen conditions at 65 bar and 600 °C, we achieved an impressive ammonia production rate of over 10 $mmol_{NH3}/(g_{cat}h)$. These results are proof of the efficiency and potential of our approach.

Key to our approach is the development of a customised reactor equipped with an alternating magnetic field for catalyst heating. The catalyst at the heart of our system consists of a ferromagnetic alloy of cobalt (Co) and nickel (Ni) surrounded by an alumina shell and decorated with 0.8 wt% ruthenium (Ru). This novel catalyst design improves reaction kinetics and ensures selectivity of ammonia synthesis, which is a significant difference from conventional catalytic systems.

Our catalysts represent a significant advance in ammonia synthesis. They are characterised by an exceptionally fast heating capability, which enables us to reach temperatures of over 800°C in less than 5 minutes. This fast temperature behaviour is crucial for the precise control of catalytic reactions and the improvement of overall efficiency. In addition, the composition of the ferromagnetic core in our catalysts is tunable, giving us the flexibility to modulate the Curie temperature of the material. This tunability allows us to optimise the performance of the catalyst under different conditions, adding a new dimension to our innovative approach. By tuning the material properties to the specific reaction requirements, we open up a range of possibilities for fine-tuning the synthesis process and improving the sustainability of ammonia production overall.

To investigate the temperature-dependent formation of the CoNi alloy, we performed temperature-programmed X-ray diffraction (XRD) analysis. Our results show that the maximum achievable temperature for heat treatment of the samples is 400 °C in an air atmosphere. Beyond this threshold, we observe the formation of Ni and Co aluminate phases together with the thermodynamically stable form of alumina, α-Al_2O_3. This means that homogeneous distribution and the formation of a CoNi alloy are no longer possible, as the fusion of these metals only occurs at significantly higher temperatures, usually above 850 °C.

Furthermore, when measuring in a hydrogen environment, we observed the formation of the CoNi alloy at about 500 °C, followed by the growth of nanoparticles (NPs) at temperatures of 850 °C and above. Despite the XRD results, H_2-TPR (hydrogen temperature programmed reduction) measurements and VSM (vibratory sample magnetometry) data suggest that reduction at temperatures below 850 °C is not sufficient to achieve the desired results.

We performed hydrogen temperature-programmed reduction to investigate the reduction profiles of bimetallic nanoparticles. The aim was to determine the optimal reduction temperature of 850 °C and to find out whether it ensures complete reduction of CoNi nanoparticles. In addition to varying the ratio of Co:Ni and CoNi:Al_2O_3 in the samples, samples with pure Co and pure Ni NPs embedded in alumina were also prepared. The latter samples should provide information on how the alloying of the metals influences the reduction profiles.

Our observations revealed a common pattern in all the materials we processed, characterised by two distinct clusters of reduction peaks. The low temperature (Low T) cluster generally ended before 400 °C, while the high temperature (High T) cluster ended before 900 °C. Remarkably, these observed reduction temperatures exceeded expectations for Co and Ni NPs. This discrepancy can be attributed to strong metal-support interactions (SMSI), a phenomenon that is well documented in the literature.

Our results suggest that Co-NPs require a much higher reduction temperature compared to Ni-NPs. Interestingly, alloying these metals lowered the temperature required for complete reduction by almost 100 °C, from 950 °C for pure Co nanoparticles to less than 850 °C for the $Co_{67}Ni_{33}$-A50 material.

We hypothesise that the low-T peak cluster is associated with CoNi-NPs that are not completely covered by alumina, leading to weak interactions. In contrast, the high-T peak cluster likely represents NPs that are closely associated with alumina and possibly fully embedded in the alumina matrix. Within the high-T cluster, the peaks below 650 °C are associated with the initial reduction step of Ni^{2+} and $Co^{2+/3+}$. The peak at about 650-700 °C probably corresponds to the complete reduction of Ni, while the peak between 750 and 800 °C is probably related to the complete reduction of Co.

Interestingly, our experiments with pure CoNi precipitate (without Al_2O_3) showed complete reduction at temperatures below 500 °C. This result is consistent with our hypothesis that the increased reduction temperature is caused by SMSI. This conclusion is further strengthened when comparing the $Co_{67}Ni_{33}$-A70 and -A50 materials, where the peaks of the high reduction temperatures are up to 70 °C lower in the latter material, mainly due to the lower alumina content and the less importance of SMSI interactions.

We investigated the magnetic properties of the synthesised materials using two vibrating sample magnetometers. The measurements were performed in air at room temperature and in an H2 atmosphere at elevated temperatures. It is found that increasing the Co content in the alloys leads to higher values of Ms (saturation magnetisation), Mr (residual magnetisation) and Hci (coercive field), independent of the VSM used. In particular, the difference in magnetisation (M), measured at 100 °C and 500 °C with an applied field of 400 Oe, becomes less pronounced with higher Co content.

X-ray photoelectron spectroscopy (XPS) was used to determine the oxidation states of the metals in the pCN-BZ and CxN1-x-AZ-850-1 particles. The focus was on Co $2p_{3/2}$ and Ni $2p_{3/2}$. We performed peak deconvolution according to the spectral fitting parameters established by Biesinger et al. (10.1016/j.apsusc.2010.10.051).

From the primary peak at 856.0 eV, it is evident that the oxidised Ni species is $Ni(OH)_2$. In the case of Co, the predominant surface species prior to reduction appears to be Co_3O_4, as indicated by the triple peak at about 780.3, 781.6 and 782.9 eV. After reduction and subsequent reoxidation, $Co(OH)_2$ is the predominant species, as evidenced by the formation of a distinct peak at about 786.5 eV. The coexistence of the two metals leads to a slight increase in the binding energy of Co by about 0.4 eV, while Ni remains relatively unaffected. In particular, the presence of alumina appears to shift the peak positions to slightly lower binding energies, which are about 0.4 eV lower.

When exposed to environmental conditions, the materials undergo a reoxidation process. During this phase, we observe some accumulation of cobalt on the surface of the nanoparticles (NP). However, in all cases we can detect traces of metallic Ni (peak at 853.4 eV) and Co (peak at 778.2 eV). It is worth noting that the amount of metal detected is consistently less than 10% in all cases.

Comparing the theoretically calculated Ms values with the measured results, significant differences can be seen, which are primarily due to the rapid passivation of the metal. As expected, no significant difference is found between the results obtained at RT in different atmospheres, as the reduction at room temperature is minimal. However, notable differences become apparent when the samples are heated to 400 °C in an H2 atmosphere and subsequently cooled, resulting in increased Ms values for all samples.

Our research highlights the feasibility and potential of demand-controlled ammonia synthesis as a key element of a sustainable energy infrastructure. By exploiting the hydrogen storage properties of ammonia and its ease of handling, combined with a customised reactor that uses induction heating and a high-performance catalyst, we present a practical solution for volatile renewable energy sources. This work contributes to ongoing efforts to advance the field of sustainable energy conversion and storage and paves the way for a cleaner and more reliable energy future.

114c Enhanced Power Output through Charged Nanopores

Johnson Dhanasekaran, Juan J. de Pablo

Charged nanochannels are promising green energy candidates and their operation is investigated in this study. Regimes of high performance are identified by spanning the large parameter space in driving force, impurities, and geometry. In a simple channel when driving forces are coupled, for example an electric potential and salt concentration gradient, a synergistic increase in electric current is reported. Using a Janus nanopore consisting of regions with different diameter and surface charge can improve performance relative to a simple geometry and this effect is shown to be highly dependent on characteristic salt concentration and choice of driving force. Coupling complex geometry with divalent impurity provides a third route to enhanced output. Calculations over the large design space are performed using the coupling of Poisson electrostatics, Nernst-Planck transport, and Navier-Stokes hydrodynamics. The continuum equations have been validated against full atomistic simulations for a sample Janus nanopore.

183a Transportation Risks and Their Mitigation Methods for Energy Transition Molecules Such As CO_2, NH_3, and H_2

William Giang, Tekin Kunt

There are various molecules of interest in energy transition. These molecules are, potentially, carbon dioxide (CO_2), ammonia (NH_3), hydrogen (H_2), etc. Safe transportation of produced CO_2, NH_3, and H_2 becomes extremely important for an efficient energy transition in the industry. The purpose of this presentation is to introduce hazards related with CO_2/NH_3/H_2 handling, its transportation through pipelines/trucks/vessels, and its storage systems. Possible mitigation methods will also be introduced as part of the presentation.

183b Role of Impurities in Transportation for Energy Systems and Their By-Products

Birol Dindoruk, Ram Ratnakar

Every energy system has its own side effects and their by-products. Disposal of by-products and carrying energy becomes extremely important from a systems engineering perspective as well as for transportation concerns. The purpose of this talk is to present various challenges in new design and repurposing of existing infrastructure in light of energy transition.

183c Critical Minerals Mining: Slurry Transport in Pipelines

Aditya Hariprasad, Hariprasad J. Subramani

Slurry pipelines are special in industrial applications which require transportation of solid particles through a pipeline. Transport of the solid particles is possible by mixing and suspending them in water or other carrier fluids. Slurry pipelines are cost-effective for long distances, reliable, and require low maintenance, low air & noise pollution. Importantly it is eco-friendly and less impact on environment compared to other transportation methods like trucking, railway, etc. Several minerals (some are critical!) have been successfully transported over long distances via slurry pipelines such as coal, iron concentrate, phosphate concentrate, copper concentrate, zinc concentrate, lead concentrate, nickel concentrate, and limestone.

The objective of this project is to determine minimum flow rate needed to transport solid particles of various types, densities, and sizes (say, sand, gravel, metal particles, etc.) and benchmark semi-empirical models available in literature for slurry transport of critical minerals.

74a Are Power-to-Heat Technologies Ready to Transition Your Facility?

Achim Wechsung, Georgios Dimitrakopoulos, Khaled Hashad, Timothy Myers

Power-to-heat generally refers to technical methods for converting electricity to some form of thermal energy that can be utilized in residential, commercial, or industrial applications. Today, many industrial processes rely on fossil fuels to generate process heat or steam. As climate change becomes a more pressing concern, societal, financial and regulatory pressure is increasing to look for alternatives. As facility owners and operators chart their path to reduced carbon intensity of their plants, they are investigating the promise and potential limitations of power-to-heat technologies.

In this presentation, we will specifically look at two common industrial processes: steam generation and direct process heating. First, we discuss three different solutions for steam generation: electrode boilers, submersible heating elements, and heat pumps. We review their operating principle, discuss typical applications, and describe limitations.

Second, we turn to direct process heating. Typical applications of these include fired-heaters, furnaces, kilns or similar equipment. The purpose is to either pre-heat a fluid before entering a subsequent processing step or directly drive a reaction. We discuss different technical solutions ranging from direct electrical heating to the generation of alternative fuels to plasma generation. We discuss their technical readiness and describe potential challenges in their deployment.

74b Maximizing Amine Contactor Efficiency with Multipoint Temperature Systems

Alina Green

Amine units are systems that treat, or “sweeten”, refinery sour gas or raw natural gas by capturing and removing unwanted compounds, primarily hydrogen sulfide (H_2S) and carbon dioxide (CO_2). An optimized gas sweetening process is essential for a plant’s overall efficiency and profitability.

Temperature monitoring and control are necessary for operating amine units, as that is how refineries and chemical plants can prevent some of the major issues seen in amine contactor columns. Reliable temperature information and a complete temperature profile can help operators optimize their units’ performance by pinpointing where most of the heat of absorption is occurring in the column. Using this information, they can minimize and even avoid foaming, carryover, corrosion, inefficient amine filtering, and other problems.

A typical contactor column has around 20 trays stacked at an equal distance from each other. A single thermocouple placed somewhere along its height provides but one data point, whereas a multipoint temperature system offers a complete temperature profile of the column. WIKA designed, manufactured, and installed multipoint thermocouple assemblies specifically for two of Gulf Coast refineries. This temperature measurement solution helped the company gain insights that allow for proactive responses to avoid operational upsets. In addition, the data creates opportunities to improve energy efficiency through optimized operation.

74c Developing a Hydrogen Training Program for Effective Change Management and Tech Integration in Energy Transition

Desiree Gajonera, Shane Finneran, Ahmed Nossair, David Burns, Amy Kitson

US and Canada have developed strategies to grow their respective hydrogen sectors. The US federal government announced an investment of $7 billion USD for 7 hydrogen hubs which emphasizes the need for workforce training to support hydrogen's integration into the existing energy market.

The hydrogen industry's progress is being driven by capital investment and government policy. Equally important yet often underestimated is the development of a skilled workforce to build and maintain this new energy sector. As the hydrogen market grows, developing a well-trained labor force will be important to oversee operation of these assets safely and effectively. This paper encourages leveraging the industry's existing talent as the most effective strategy. The energy transition faces significant challenges which highlights the urgency for a comprehensive Hydrogen Training and Certification Program.

This paper explores the challenges of transitioning the existing labor force for the hydrogen industry. Given this sector's early phase, the paper applies a comprehensive approach to asset management from project initiation to operation. The study investigates the full hydrogen value chain, including (1) Hydrogen Production, (2) Storage, Transmission, and Distribution, and (3) Application. In collaboration with gas operators in North America, a gap analysis for each area was performed to determine the necessary skills and competencies. The paper proposes a strategic roadmap that focuses on essential areas to overcome these challenges and develop an in-depth hydrogen training and competency program.

95a Technical Safety Studies/ESG for SAF Capital Projects

Alexandre Cardoso, Kierra McDougall

With the continued projected demand for aviation fuel, transitioning to Sustainable aviation fuels (SAFs) is a smart move toward meeting the Fly Net Zero Plan, to achieve net zero carbon emissions in the aviation industry by 2050. The International Air Transport Association (IATA) is leading the Fly Net Zero Plan, in alignment with the Paris Agreement. These novel fuel types are a promising renewable energy alternative to traditional fossil jet fuels, as they produce 85% less carbon emissions than traditional jet fuels over their life cycle and are compatible with current air transportation infrastructure.

SAF production processes are relatively new, so there is little available information on prevalent process hazards and safety concerns for such fuels. However, since high flammability and toxic consequences are already associated with current aviation fuel, similar properties are anticipated for SAF processing and final products. For proposed SAF capital projects, it is therefore imperative that technical safety studies are performed during the investment, planning, designing, construction, and commissioning stages to identify and manage business risks including impacts on the environment, employees, and local communities.

Each SAF process is unique in its feedstock, intermediates, final product, and steps required for the overall process. This presentation will identify several different SAF production technologies, with a brief description of each process, resultant product properties, and the expected potential hazards.

Process safety is one of the many focuses during a capital project and emphasizes understanding the associated hazards of the process throughout the different phases that make up the Project Life Cycle, covering the project from onset to completion. Despite the potential process safety hazards associated with SAF production and use, this low-carbon technology is an ideal candidate for funding due to its alignment with Environmental, Social, and Governance (ESG) strategies. Incorporating technical safety into SAF start-up companies and projects will further demonstrate the significant contribution of SAF technology to ESG strategies beyond a carbon footprint focus.

Appropriate safety studies must be conducted to the level of detail best suited to the available project data throughout the process. This presentation will identify three key technical safety studies that can impact a company's ESG strategy for a given capital project for SAF production: Process Hazards Analysis (PHA), Consequence and Risk Based Facility Siting (FSS), and Insurance Risk Engineering (IRE). Each technical safety study and its impact on ESG framework will be further discussed.

95b Methanol Economy: Energy Transition and Decorbonising India

Divesh Kumar

Topical 15: The Energy Transition

METHANOL ECONOMY: ENERGY TRANSITION & DECORBONISING INDIA

Energy plays a crucial role and is considered as key indicator of economic development & overall progress of any country. Timely and sufficient availability of high-quality energy at an affordable price is of great importance, as it ensures the operation of the economy and leads to sustainable growth. Majority of the countries are dependent on fossil fuels for their energy demand and are seized with the problem of establishing a balance between environment & technologies to minimize the impact of fossil fuel on the environment. Sheer size of developing countries in terms of size and population majority being in Asia & Africa coupled with huge scope for growth means their energy demand is set to grow by more than that of any other part of the world in the coming decades.

Various countries commitment to achieving net-zero emissions as stated at Conference of Parties (COP), necessitates a transition to clean energy. Interruption in energy or failure to meet the increase in energy demand leads to bottleneck in economy. With soaring crude oil demand and import bills draining foreign exchange, the need for an alternative reliable fuel is imperative to enhance economic stability and growth of developing countries.

Reasons outlined above signals for development reliable fuel of global relevance for shifting to alternate cleaner fuels to fulfil commitments of net zero, growing energy demand and environmentally sustainable development.

ENERGY TRANSITION ROADMAP FOR INDIA

Indian govt. has prepared a robust roadmap for energy transition, few of them are increasing power generation capacity from renewable energy like solar and wind, unveiling National Green Hydrogen Mission for reduction in dependence on imported fossil fuels and feedstock, National Policy on Bio Fuels etc. Further, final roadmap for 'Methanol Economy' being worked out by National Institution for Transforming India (NITI) Aayog and will be unveiled soon.

ENERGY TRANSITION & DECARBONISATION: METHANOL AS A FUEL

Methanol (CH3OH) is a liquid alcohol fuel that can be synthesized from various feedstocks, including natural gas, coal, biomass, and carbon dioxide. The emergence of affordable renewable energy sources such as wind, solar, geothermal, and hydropower has enabled the production of 'green methanol,' which is environmentally friendly and significantly reduces greenhouse gas emissions.

Methanol plays crucial role in chemical industry as a base material for manufacturing wide range of products such as polymer fibres for the textile industry, plastics for packaging, glues, adhesives, diapers, paints, and various solvents among others. Additionally, methanol can be utilized as a fuel in the transportation sector, either by blending it with gasoline or diesel fuels with minor engine modifications or as a standalone fuel. Green methanol is a cleaner and more sustainable choice for both industrial and transportation purposes, contributing to the reduction of greenhouse gas emissions and air pollutants.

Different volumes of methanol blends such as M5, M15, M85 or M100 can be used depending upon the application. China world's 2nd largest economy, now uses methanol as a vehicle fuel in blends ranging from five percent to 100 percent ('M5' to 'M100'), with a view to take-up of entirely methanol-powered vehicles.

The Methanol Economy is a concept that advocates using methanol as a versatile and efficient energy carrier to promote greener and more sustainable future. It also facilitates the development of hydrogen economies and thus has gained attention as a potential alternative to traditional fossil fuels, addressing energy and environmental challenges. Methanol holds strategic importance in India's efforts to achieve decarbonization and enhance energy security.

FUTURE ROADMAP FOR METHANOL ECONOMY

- Production of methanol from Indian high ash coal from indigenous technology.
- Production of methanol through carbon capture technology.
- Bio-mass, Stranded Gas & MSW for methanol production.
- Utilization of methanol & DME in transport sector like railways, road, marine and defence, Industrial Boilers, Diesel Gensets & Power generation & Mobile towers etc.
- Utilization of methanol and DME as domestic cooking fuel- cook stoves.
- Utilization of methanol in fuel cell applications in Marine, Gensets and Transportation

This study explores the pivotal role of methanol in driving India's energy transition and advancing decarbonization initiatives. It examines the multifaceted benefits of methanol as a clean energy source, emphasizing its potential to significantly reduce greenhouse gas emissions and promote sustainable energy practices. Additionally, the paper underscores the importance of reducing India's reliance on fossil fuels, particularly in the context of achieving energy security and contributing to global efforts for a cleaner & more sustainable future.

95c A Methodological Sustainability Assessment to Process Intensification (MSAtoPI) By Reactive-Separation Systems

Seckin Karagoz

To produce clean and cheap power by eliminating the economic and energy penalties, it is necessary to develop new technologies or improve the existing processes based on sustainability criteria and alternative fuels. Integration of sustainability into process design contributes to the achievement of this target by eliminating or minimizing negative effects. However, measuring sustainability performance and decision-making are crucial and challenging steps for the development of sustainable industrial processes. The framework for Methodological Sustainability Assessment to Process Intensification (MSAtoPI) is thus introduced in this study as a methodological approach with appropriate sustainability indicators that ensures the capability of gathering and abstracting the complex process operations (energy, mass, and momentum transport phenomena), and that offers clear analysis and communication. The main concepts are demonstrated in practice by using a biomass-based, highly efficient, and ultra-compact process to produce H_2 while simultaneously capturing CO_2 in an IGCC (Integrated Gasification Combined Cycle) power plant, which is intensified by reactive-separation systems (a hybrid membrane reactor (MR)/adsorptive reactor (AR) system). MSAtoPI clearly concretizes that the proposed intensified reactive-separation systems have huge potential to produce clean and cheap power by eliminating the economic and energy penalties by having more sustainable natures (minimum: ~ 18%, average: ~ 75%, maximum: ~140% improvement in sustainability).

Hydrogen Separation and Compression Using an Eco-Friendly Electrochemical Hydrogen Pump to Decarbonize the Economy

Karthik Arunagiri[1]

[1] Department of Chemical Engineering, The Pennsylvania State University, University Park, PA, 16802, USA, kva5538@psu.edu

Abstract: Hydrogen storage and transportation on a large scale is a bottleneck for decarbonization and the development of a hydrogen-based economy. US DOE has proposed to use the existing natural gas pipeline for the storage (at low concentration due to embrittlement problem) and distribution of hydrogen. To separate hydrogen from natural gas at the place of use a continuous and stable process such as an electrochemical hydrogen pump (EHP) is used. Novel materials such as ion-pair high-temperature polymer electrolyte membranes (HT-PEMs) paired with phosphonic acid ionomer electrode binders have substantially improved the performance of HT-PEM EHPs and fuel cells. Our recent work of blending poly(pentafluorstyrene-co-tetrafluorostyrene phosphonic acid) (PTFSPA) with Aquivion® improved ionomer conductivity, gas diffusivity, and kinetics resulting in a remarkable EHP performance of 5.1 $A.cm^{-2}$ at 0.4 V at T = 200 °C (under pure hydrogen) when used as an electrode binder. In this work, 3 types of phosphonic acid ionomers, poly(vinyl phosphonic acid), poly(vinyl benzyl phosphonic acid) by themselves and when blended with Aquivion® were studied as binders. These studies were performed in the context of an EHP platform – both membrane electrode assemblies (MEAs) and interdigitated electrode arrays (IDAs) decorated with nanoscale platinum electrocatalysts with thin film ionomer electrolytes. For all phosphoric acid ionomer types, the addition of Aquivion® promoted ionic conductivity, hydrogen oxidation/evolution reaction kinetics (HOR/HER), and hydrogen gas permeability. Furthermore, the best-performing binder blend of PTFSPA-Aquivion® was shortlisted for further EHP study. In this study, hydrogen from a natural gas mixture (at various concentrations) was separated using EHP at various temperatures (200 °C, 160 °C, and 120 °C). Furthermore, the effect of back pressure on the anode exit and the cathode exit was studied to assess the performance of the EHP to compress hydrogen. The exit hydrogen purity was analyzed using micro-GC. With this study, we were able to demonstrate a new electrochemical technology for purifying and compressing hydrogen.

Introduction and Background

Hydrogen will play a central role in decarbonizing the global economy in the coming years. About 10 % of carbon dioxide (CO_2) emissions hail from steel and ammonia manufacturing [1, 2]. Green hydrogen can be used as an alternative reducing agent in steel manufacturing as opposed to carbon monoxide (CO) derived from fossil fuels[3]. Furthermore, hydrogen has the potential to be a cost-effective energy storage medium for long-term/seasonal energy storage[4, 5] and the fuel of choice for heavy-duty vehicle (HDV) transport[6]. To make the hydrogen economy viable it is necessary to reduce the cost of hydrogen. United States DOE has the target of reducing the cost of 1kg of hydrogen to 1 USD in 1 decade[7]. In this regard, it is necessary to reduce the storage and transportation cost, which accounts for 50% of the total cost of hydrogen. One of the solutions is to use the existing natural gas pipeline infrastructure to store and distribute hydrogen. One of the key issues of blending hydrogen in natural gas pipelines is the embrittlement of pipeline materials with hydrogen This concern is alleviated by diluting hydrogen with natural gas to a value of 20% less, preferably 3 to 8%. As shown in **Figure 1,** the process would use the hydrogen generated by water electrolysis rather than the steam reforming of methane (95% of currently produced hydrogen). The energy required for electrolysis and EHP would be supplied from renewable solar and wind energy. This would also help in the load-shedding of renewable energy. Electrochemical hydrogen pump (EHP) can be used to separate hydrogen. However, natural gas has sulfur components that may affect the performance of the EHP. In a typical EHP, hydrogen along with natural gas fed continuously to the device at its anode side is oxidized electrochemically in the presence of a catalyst to produce protons, which are transported

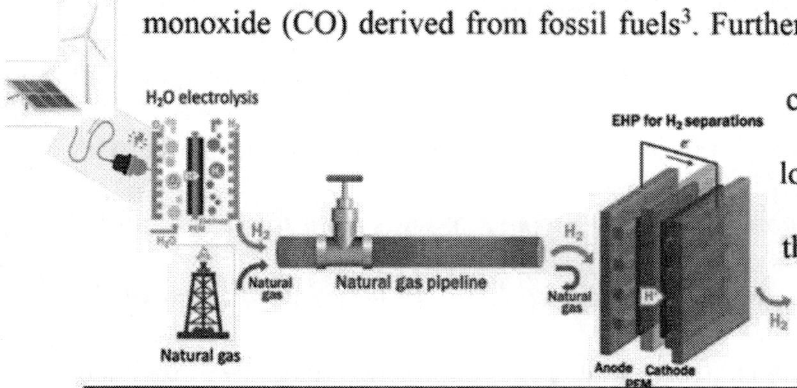

Figure 1. Importance of EHP in purify hydrogen for the green hydrogen economy

to the cathode side by the proton-conducting polymer electrolyte. The protons react at the cathode (accepting the electrons released from the anode) to produce hydrogen. The process is electrolysis and requires energy given to the system. The amount of electrical energy that is required to operate the cell is dependent upon minimizing the losses incurred from electrochemical reaction kinetics, proton transport through the membrane, and the delivery of reactants to active reacting sites. The voltage losses from these sources are described as overpotentials. The activation overpotential corresponds to reaction kinetics, while the ohmic overpotential is inversely commensurate with ionic and electrical conductivity and the concentration overpotential hails from the mass transfer limitation of reactants and product. The present work will enable a systematic reduction of the resistances in EHP and reduce the platinum catalyst loading. It is envisioned that new knowledge about the catalyst-ionomer-other impurities interaction would be important for realizing new EHP electrodes with lower PGM loadings.

Electrochemical hydrogen pump (EHP) is a compelling technology for purifying the hydrogen from the natural gas-hydrogen mixture in addition to compressing the purified hydrogen simultaneously. The first generation of EHPs[8] by General Electric used perfluorosulfonic acid (PFSA) PEMs, and they operated at low temperatures (< 100 °C) as the PFSA PEMs need condense water to mediate proton conductivity. Neyerlin et. al. used lower temperature hydrogen pump[9] with Nafion™ membrane to study the exchange current density of hydrogen oxidation and evolution reaction. The EHPs based on Nafion™ are restricted by operating temperature range (below 110 °C) as the membrane losses conductivity at higher temperatures. The low-temperature EHP suffers from average performance and catalyst poisoning by impurities from the reactant gas mixture (CO and other gases). To overcome this problem, high temperature EHPs (till 250 °C) have been used[10, 11]. High temperature EHPs use phosphoric acid doped ion-pair membrane[12] and poly(tetrafluorostyrene phosphonic acid) as binder[11]. Phosphoric acid-based ionomer can conduct at higher temperatures as it does not require an external water molecule. This system can be used to separate hydrogen gas having other impurities.

Figure 2a shows a schematic working of EHP. Hydrogen mixed with other gases (moslty) enters the anode side of the cell wherein hydrogen oxidation reaction happens and protons, electrons are produced. Protons cross through the membrane and electrons go through the external circuit. All the gases are repelled back to the anond exhaust. Hydrogen evolution reaction happens on the other side of the cell in the cathode where protons combines with electrons to produce hydrogen. **Figure 2b** gives a zoomed-in view of a model cathode structure that accentuates the electrode ionomer binder in HT-PEM electrodes. Notably, the protons and electrons for the HOR and HER reactions need to intersect at the platinum electrocatalyst – which can be partially covered by the electrode binder. The minimum cell voltage for the EHP if the fugacity of hydrogen on the anode and cathode are the same is 0 V. However, the practical operation of the EHP still necessitates applied potential due to various overpotentials related to reaction kinetics, ohmic drops, and gas mass transfer resistances. The binder is an ionomer that coats the catalyst and provides mechanical stability to the electrode catalyst layer as shown in **Figure 2b**. The binder invariably affects the ohmic, activation, and gas transport properties of the electrode layer. The

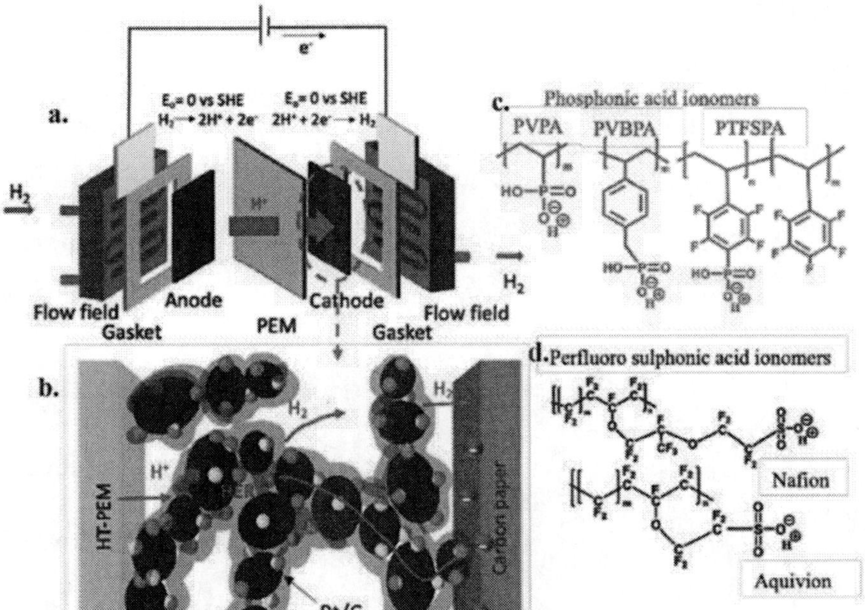

Figure 2. a) Electrochemical hydrogen pump (EHP) (b) Electrode material and processes involved in the electrodes i.e., gas transport, electrochemical reaction, and ion transfer to emphasize the importance of binder; (c) and (d) Chemical structures of binder materials

electrode binder has a profound impact on the performance of electrochemical devices such as EHP[11] and fuel cells[13]. Hence, the binder has a profound effect on the performance of EHPs. It is necessary to study the influence of the binder on the electrode to better understand the system.

EHPs can be used to separate hydrogen from other gas mixtures. Gokul et al., 2022 used the HT-PEM EHP system to separate hydrogen from carbon monoxide (CO) gas mixtures[10] at various concentrations. The highlight of this work is that CO was used in the anode side of EHP at high concentrations and hydrogen at lower concentrations at higher temperatures. This was possible due to the usage of ion-pair HT-PEM and PTFSPA binder (**Figure 3a**) which conducts at higher temperatures around 200 °C. The development of phosphoric acid imbibed polycation HT-PEM (known as ion-pair HT-PEM) by Arges Lab[12] and Los Almos[14] has overcome the limitation of conventional PBI-based HT-PEM membranes. The PBI-based membranes have the problem of phosphoric acid evaporation and loss at higher temperatures at 200 °C. The ion-pair based membrane has an electrostatic attraction with the phosphate anion from phosphoric acid. This leads the membrane to retain phosphoric acid at higher temperatures and prevents evaporation. Paring the ion-pair based membrane with PTFSPA binder has been found to improve the performance of EHP. The PTFSPA has the phosphonic acid tethered to the polymer backbone. Further, the fluorine in the styrene ring is a high electronegative

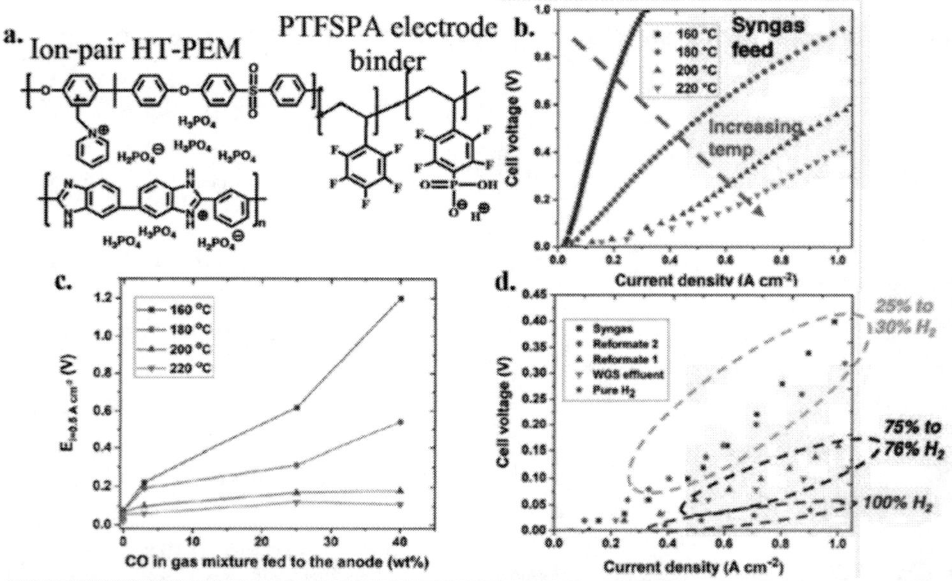

Figure 3: (a) Chemical Structure of ion-pair HT-PEM and PTFSPA electrode binder; (b) Polarization curve of EHP at different temperatures with synthesis gas feed; (c) Effect of CO on the voltage at 0.5 A.cm^{-2}; (d) Effect of hydrogen concentration at the anode of EHP (Gokul et al., 2022)

group which makes the phosphonic acid more acidic in nature. This leads to improved conductivity and reduced anhydride formation at higher temperatures[13]. However, as shown in Table 1, PTFSPA still forms anhydride and the addition of PTFSPA with Aquivion® leads to no anhydride formation, and when used as a binder improves the performance of EHP as shown in **Figure 3b**. Gokul et al., 2022 studied the effect of CO concentration and hydrogen concentration on EHP polarization. As shown in **Figure 3c**, at higher temperatures the EHP performance was better. At higher temperatures, the polarization improved which suggests that the adsorption (poisoning) of CO on the platinum catalyst is reduced with an increase in temperature. Figure 4.1 c shows the effect of CO concentration on the voltage required for $0.5 A.cm^{-2}$. At 200 °C the concentration dependence was negligible. Also, the polarization changes as a function of hydrogen concentration are shown in **Figure 3d**.

Even though the effect of carbon monoxide on EHP polarization was studied. The effect of other gaseous components such as hydrocarbons, etc. (present in the natural gas mixture). is still unknown. It is necessary to study the effect of them on the EHP performance, as the sulfide components are known to poison the platinum catalyst and deteriorate the performance of the system[15]. The proposed work plan to address the following knowledge gap:

- The effect of gas components (entering the anode side of the EHP) hydrocarbons, etc. on the platinum catalyst and the performance of EHP is not known.
- Even though the performance of HT-PEM EHP is better compared to that of the LT-PEM EHP, the platinum loading in the HT-PEM system is still very high (2 mg_{Pt} cm^{-2} for the MEA). At lower loading, the limiting current density values were reached[11].

Results & Discussion

Initial experiments were to study the interaction of the ionomer-catalyst behavior and the effect of the ionomer on the EHP performance. The ionomer studied comprises blend of one of three phosphonic acid ionomers (PVBPA, PTFSPA and PVPA) with one of two sulphonic acid ionomers (Nafion™ and Aquivion®) as shown in **Figure 1 and Table 1**. Poly(vinyl benzyl phosphonic acid) PVBPA was prepared by phosphonating commercially available poly(vinyl benzyl chloride) (PVBCl) using triethyl phosphite and following the Arbuzov reaction[16] procedure. Poly(tetrafluorostyrene phosphonic acid) (PTFSPA) was prepared by phosphonating the polymer Poly(pentafluorostyrene)[13]. To ensure the polymer has adequately high number average molecular weight (M_n) and to have uniform properties, we have synthesized poly(pentafluorostyrene) by emulsion polymerization in our lab. We attained 90% conversion of pentafluorostyrene monomer during the reaction and M_n of 68 kDa. Poly (vinyl phosphonic acid) polymer and sulphonated polymer were procured commercially.

Table 1. Phosphonic acid and sulfonic acid ionomer blends and their electrochemical properties

Phosphonic acid ionomer and IEC value(mmol g-1)[a]	Perfluorosulfonic acid ionomer and IEC value (mmol g^{-1})[b]	P/S ratios studied	Anhydride[c] formation for pure ionomer/ and its blend with Aquivion® (%)	Maximum thin-film conductivity [d] for pure ionomer/ and its blend with Aquivion® ($mS.cm^{-1}$)
PTFSPA (2.6)	Nafion™ (0.91)	0.5, 1, 1.6, 2, 4.7	21.2/0	18.2/38.5
	Aquivion® (1.02)			18.2/42.5
PVPA (7.8)	Aquivion® (1.02)	1, 2, 4, 5.3	58.3/0	12.3/34.5
PVBPA (6.23)	Aquivion® (1.02)	2.7	100/48	1.2/8.9

[a, b] Either measured using acid-base titration or calculated from equivalent weight; [c] Evaluated by solid-state NMR ; [d] Evaluated by thin-film coated on interdigitated electrode

The synthesized ionomers were tested for their electrochemical properties. The ionomer conductivity was tested by spin-coating it as a thin film on an interdigitated electrode platform. Interdigitated electrode (IDEs) was used to measure the thin-film electrochemical properties of the ionomers and their blends. In the electrodes of a fuel cell, the ionomer binder holds the electrocatalyst particles together. As a result of this, they coat the surface of the electrocatalyst nanoparticle and electrocatalyst support as thin films. The IDEs allow our team to evaluate the electrochemical properties of thin film ionomers to understand their electrochemical properties as thin films. The conductivities of the ionomers are listed in **Table 1**. Equation 1 below was used to measure proton conductivity (κ).

$$\kappa = \frac{D}{R_{film} \cdot l \cdot t_{film} \cdot (n-1)} \qquad (1)$$

In this equation, D is the distance between the teeth, l is the length of the teeth, and n is the number of teeth in the IDE. R_{film} is the resistance of the film obtained from electrochemical impedance spectroscopy (EIS) and t_{film} is the thickness of the film obtained from ellipsometry. The thickness of the film was determined by the ellipsometry technique. As seen in **Table 1**, the blend of PTFSPA with Nafion™ or Aquivion® show higher conductivity than PTFSPA by itself. Out of the two PFSAs (Nafion™ or Aquivion®) studied, Aquivion® with PTFSPA showed the highest proton conductivity. After this, we decided to only blend Aquivion® with other phosphonic acid ionomer materials (i.e., PVPA and PVBPA) in addition to PTFSPA. It is interesting to note that PVPA has very less conductivity and it did not conduct at higher temperatures. Further, when PVPA is blended with Aquivion® the conductivity improves, and it conducts at higher temperatures. This led us to study the behavior of the ionomers at a higher temperature (around 200 °C). It is well known the that the phosphonic acid ionomers forms anhydride at higher temperature[13, 17]. This leads to a reduction in the conductivity of the ionomers. Anhydride formation of the various phosphonic acid ionomers and phosphonic acid ionomers blended with Aquivion® was monitored via solid-state ^{31}P NMR. The chemical shift in ^{31}P NMR spectra was used to identify anhydride formation and integration of the peaks that corresponded to phosphonic acid and phosphate ester (i.e., anhydride) allowed us to determine the extent of anhydride formation. The results of the anhydride formation are summarized in **Table 1**. It is seen that with the addition of Aquivion® all the phosphonic acid ionomers have less anhydride formation and improved thin-film conductivity.

Apart from ionic conductivity and its impact on ohmic resistances in electrodes, it is necessary to test how the said ionomers and ionomer materials affect electrode kinetics and gas permeability. Hence, we ran MEA polarization curves with electrodes made from 4-types of binders: 1.) PTFSPA, 2.) PTFSPA blended Aquivion®, 3.) PVPA, and 4.) PVPA blended with Aquivion®.These MEA EHP study deployed the same type of an ion-pair high-temperature proton exchange membrane (HT-PEM), catalyst (Pt/C – 37% Pt on carbon by TKK) [12] and catalyst loading (1 mg_{Pt} cm^{-2} on each electrode). The binder loading in the electrodes was 10 wt%. The experiments were conducted at 200 °C.

The polarization curves of in **Figure 4a** shows that the ionomer blend of PTFSPA and Aquivion® have less polarization when compared to the MEA that only used PTFSPA as the electrode binder. The polarization curve exhibits two main regions: a linear region and an asymptotic region. The linear region refers to the ohmic and kinetic losses dictating the resistance in the cell. The asymptotic region refers to the limiting current region, where the system is limited by hydrogen transport in the porous electrodes. The limiting current for the MEA with PTFSPA only as the electrode binder was lower than the PTFSPA-Aquivion® as the electrode binder. Similar observations were observed with MEAs made with PVPA and PVPA blended with Aquivion® - except these systems display greater polarization. Still, blending Aquivion® with PVPA gave reasonable EHP performance - 2.3 A cm^{-2} at 650 mV. **Figure 4b** presents the Nyquist plot for MEAs tested with a background voltage of 0.05V. The small DC bias ensured that the cell was kinetically controlled and not significantly impacted by mass transfer. The charge transfer resistance (R_{ct}) obtained from the radius of the semi-circle, is lower for the blended ionomer compared to that of the phosphonic acid ionomers by themselves (both of PTFSPA and PVPA). The R_{ct} values are extracted from the modelling of the EIS data and reported in **Figure 4c**. As we can see there is a significant reduction in R_{ct} for PVPA ionomer when blended with Aquivion®. For the PTFSPA ionomer also there is a reduction in R_{ct} when blended with Aquivion®. EIS data is obtained with a DC bias voltage of 0.8 V in the background caused the MEAs to be mass transfer limited. Analyzing the Nyquist plots (not shown), we extracted the diffusion resistance by plotting the real impedance versus the inverse of square root of angular frequency (i.e., generated a Warburg plot – see **Figure 4d**. The slope of the line in the Warburg plot is the diffusion resistance (σ). σ is inversely related to the square root of the diffusion coefficient. A lower σ value means better gas diffusivity in the porous electrodes – which is affected by the binder selection. Overall, we see that blending Aquivion® with phosphonic acid ionomers promotes reaction kinetics and hydrogen gas

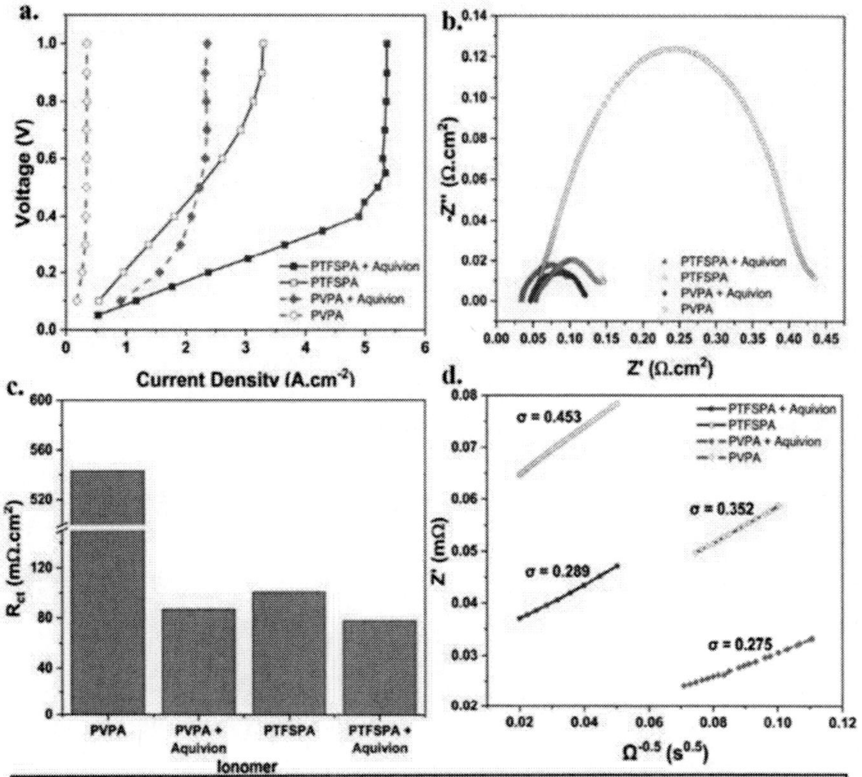

Figure 4. (a) Chronoamperometry and (b) EIS with a DC bias of 0.05V for EHP (T = 200°C) having PTFSPA and PVPA with and without Aquivion® as binders in the MEA. (c) Rct obtained from the EIS data (d) Warburg plot of Z' vs. the reciprocal of the square root of angular frequency ($\Omega^{-0.5}$).

transport in porous electrodes used in EHPs. We were able to see that the charge transfer coefficient improves when blending the phosphonic acid ionomer with the sulphonic acid ionomer. This is mainly due to the reduced adsorption of the sulphonate anion (from sulphonic acid ionomer) on the Pt catalyst than the phosphonate anion (from phosphonic acid ionomer), which was confirmed by our DFT study.

Furthermore, we used the EHP platform to separate hydrogen from a dilute hydrogen-natural gas mixture. **Figure 5a** presents the polarization curves of the EHP at 200 °C with an ion-pair HT-PEM and phosphonic acid ionomer-PFSA ionomer blend electrode binders with different dilute hydrogen-natural gas (H_2-NG) feed mixtures to the anode and a control case of pure hydrogen to the anode. No back pressure was applied to the anode or the cathode in these experiments. The pure hydrogen polarization curve data was reported in our previous work.[18] The polarization curves at different cell temperatures, 120 °C and 160 °C, were also performed. The polarization data in **Figures 5a** demonstrate that the cell polarization increases when feeding more dilute hydrogen-natural gas mixtures. We anticipated this effect for two reasons: i) the Nernst equation predicts that the minimum cell voltage needed for purifying

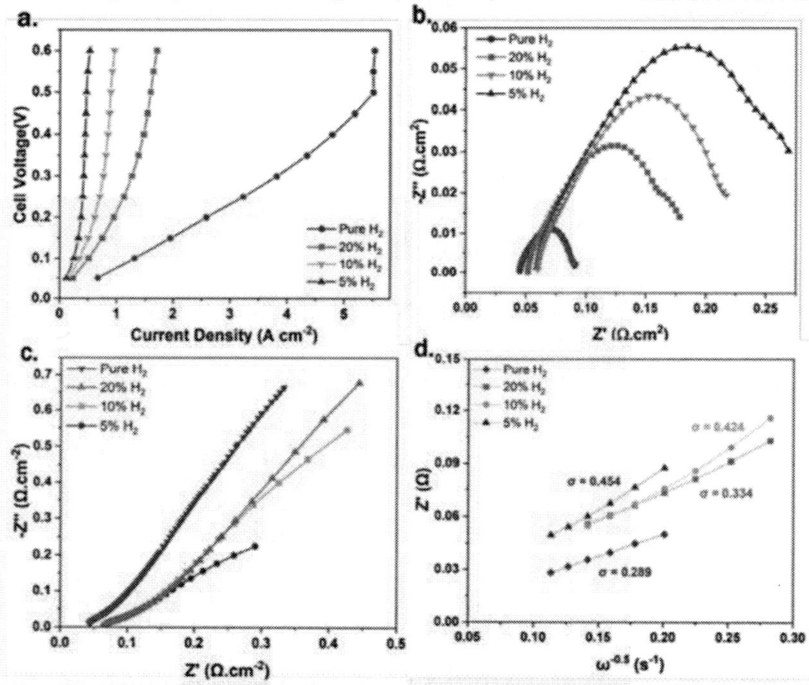

Figure 5. (a) EHP polarization curves for the different H2-NG mixtures and pure H2 at 200 °C; (b) EIS with DC bias of 0.05V at T = 200 °C with H2-NG mixtures and pure H2; (c) EIS with DC bias of 0.55V at T = 200 °C with H2-NG mixtures and pure H2; (d) Warburg plot showing the diffusion resistance (σ) calculated from the Nyquist plots in (c).

the hydrogen increases when starting with a more dilute feed on the anode side while maintaining approximately the same purity and partial pressure of hydrogen on the cathode side and ii) we anticipated greater diffusion resistance within the porous anode when feeding more dilute hydrogen. The purity of the hydrogen from the cathode of the EHP was analyzed using micro-GC. The purity was analyzed at two different current densities: one at a lower current density and other near the limiting current. The purity of the hydrogen from the cathode of EHP was $\geq$ 99.3vol% at 160 and 200 °C. At 120 °C, the purity of the hydrogen from the cathode was 99.039 to 99.096 vol%. Furthermore, we did the electrochemical impedance spectroscopy (EIS) experiment to elucidate various resistance from the cell. The results of EIS with dc bias of 0.05V (in the kinetics-controlled region) is shown in figure 5b and the EIS with dc bias of 0.55V (in the gas-transport controlled region) is shown in figure 5c. An interesting observation is that the charge transfer resistance or the kinetic resistance (semicircle diameter) increases at lower hydrogen concentration. The increase of diffusion resistance (σ) at low hydrogen concentration, as shown in **figure 5d,** is much more obvious than the kinetic resistance onservation.

The EHP experiments of this study helped us understand the interaction of the ionomer binder with the platinum catalyst used in the electrodes of EHPs. This warrants us to understand the effect of various constituents (phosphoric acids from the membrane, binder and contaminants) on the EHP performance when used to separate and compress hydrogen. Revealing this is important to enhance the electrocatalyst utilization of EHPs. From the data in **Figure 4a**, it can also be seen that the EHPs reached a limiting current (mass transfer limitation) at about 0.45 V. The future research outcome would be to understand the catalyst binder interactions and to improve the gas transport in the electrode of EHP, to improve the polarization and to improve the limiting current density. Additionally, the experimental and DFT study would help us under the catalyst poisoning occurring in the electrodes of EHP.

References:

1. Hannah Ritchie, M. R. a. P. R., CO_2 and Greenhouse Gas Emissions. *Our World in Data* **2020**.

2. Lamb, W. F.; Wiedmann, T.; Pongratz, J.; Andrew, R.; Crippa, M.; Olivier, J. G. J.; Wiedenhofer, D.; Mattioli, G.; Khourdajie, A. A.; House, J.; Pachauri, S.; Figueroa, M.; Saheb, Y.; Slade, R.; Hubacek, K.; Sun, L.; Ribeiro, S. K.; Khennas, S.; de la Rue du Can, S.; Chapungu, L.; Davis, S. J.; Bashmakov, I.; Dai, H.; Dhakal, S.; Tan, X.; Geng, Y.; Gu, B.; Minx, J., A review of trends and drivers of greenhouse gas emissions by sector from 1990 to 2018. *Environmental Research Letters* **2021,** *16* (7), 073005.

3. Bhaskar, A.; Abhishek, R.; Assadi, M.; Somehesaraei, H. N., Decarbonizing primary steel production : Techno-economic assessment of a hydrogen based green steel production plant in Norway. *Journal of Cleaner Production* **2022,** *350*, 131339.

4. Monforti Ferrario, A.; Vivas, F. J.; Segura Manzano, F.; Andújar, J. M.; Bocci, E.; Martirano, L. Hydrogen vs. Battery in the Long-term Operation. A Comparative Between Energy Management Strategies for Hybrid Renewable Microgrids *Electronics* [Online], 2020.

5. Kharel, S.; Shabani, B. Hydrogen as a Long-Term Large-Scale Energy Storage Solution to Support Renewables *Energies* [Online], 2018.

6. Pardhi, S.; Chakraborty, S.; Tran, D.-D.; El Baghdadi, M.; Wilkins, S.; Hegazy, O. A Review of Fuel Cell Powertrains for Long-Haul Heavy-Duty Vehicles: Technology, Hydrogen, Energy and Thermal Management Solutions *Energies* [Online], 2022.

7. Pivovar, B. S.; Ruth, M. F.; Myers, D. J.; Dinh, H. N., Hydrogen: Targeting $1/kg in 1 Decade. *The Electrochemical Society Interface* **2021,** *30* (4), 61.

8. Sedlak, J. M.; Austin, J. F.; LaConti, A. B., Hydrogen recovery and purification using the solid polymer electrolyte electrolysis cell. *International Journal of Hydrogen Energy* **1981,** *6* (1), 45-51.

9. Neyerlin, K. C.; Gu, W.; Jorne, J.; Gasteiger, H. A., Study of the Exchange Current Density for the Hydrogen Oxidation and Evolution Reactions. *Journal of The Electrochemical Society* **2007,** *154* (7), B631.

10. Gokul Venugopalana, E. A., Luis Briceno-Menaa, José Romagnolia, John Flakea, Christopher G. Arges, Electrochemical pumping for challenging hydrogen separations. *ACS Applied Materials (under review)* **2022**.

11. Venugopalan, G.; Bhattacharya, D.; Kole, S.; Ysidron, C.; Angelopoulou, P. P.; Sakellariou, G.; Arges, C. G., Correlating high temperature thin film ionomer electrode binder properties to hydrogen pump polarization. *Materials Advances* **2021,** *2* (13), 4228-4234.

12. Venugopalan, G.; Chang, K.; Nijoka, J.; Livingston, S.; Geise, G. M.; Arges, C. G., Stable and Highly Conductive Polycation–Polybenzimidazole Membrane Blends for Intermediate Temperature Polymer Electrolyte Membrane Fuel Cells. *ACS Applied Energy Materials* **2020,** *3* (1), 573-585.

13. Atanasov, V.; Lee, A. S.; Park, E. J.; Maurya, S.; Baca, E. D.; Fujimoto, C.; Hibbs, M.; Matanovic, I.; Kerres, J.; Kim, Y. S., Synergistically integrated phosphonated poly(pentafluorostyrene) for fuel cells. *Nat Mater* **2021,** *20* (3), 370-377.

14. Lee, K.-S.; Spendelow, J. S.; Choe, Y.-K.; Fujimoto, C.; Kim, Y. S., An operationally flexible fuel cell based on quaternary ammonium-biphosphate ion pairs. *Nature Energy* **2016,** *1* (9), 16120.

15. Sethuraman, V. A.; Weidner, J. W., Analysis of sulfur poisoning on a PEM fuel cell electrode. *Electrochimica Acta* **2010,** *55* (20), 5683-5694.

16. Kim, S. H.; Park, Y. C.; Jung, G. H.; Cho, C. G., Characterization of poly(styrene-b-vinylbenzylphosphonic acid) copolymer by titration and thermal analysis. *Macromolecular Research* **2007,** *15* (6), 587-594.

17. Bingöl, B. Synthesis and characterization of Poly(vinylphosphonic acid) for proton exchange membranes in fuel cells. Ph.D. Thesis, Mainz, 2007.

18. Arunagiri, K.; Wong, A. J.-W.; Briceno-Mena, L.; Elsayed, H. M. G. H.; Romagnoli, J. A.; Janik, M. J.; Arges, C. G., Deconvoluting charge-transfer, mass transfer, and ohmic resistances in phosphonic acid–sulfonic acid ionomer binders used in electrochemical hydrogen pumps. *Energy & Environmental Science* **2023,** *16* (12), 5916-5932.

135b Development of a CO_2 Post Combustion Capture Technology Tailored for Gas Turbine Exhaust Using Solid Sorbents

Yui Kotani

In a joint development with JERA (Japan's largest power generation company) and RITE (Research Institute of Innovative Technology for the Earth, Japanese center of excellence committed to research environmental issues), Chiyoda Corporation is developing a solid sorbent based CO_2 post combustion capture technology especially tailored for gas turbine exhaust under the fund of Japanese government (fund agent NEDO: New Energy and Industrial Technology Development Organization). Due to the low CO_2 concentration in gas turbine flue gas (3-5 vol%), although it has been technically possible to remove the CO_2 using conventional liquid amine based technology, the barrier of economic feasibility has been very high to implement such process to thermal power plants. The project aim is to halve the cost needed to remove the CO_2 from such dilute flue gas and be commercially ready by 2031.

In the project, we are scheduling a bench scale plant followed by few 10 tCO_2/day slip stream demonstration at JERA's natural gas fired combined cycled power plant along with a joint demonstration of CO_2 utilization using the captured CO_2. We are still in the midst of deciding the CO_2 utilization technology, which could pose an opportunity to demonstrate their technology in Japan.

Amine based solvents has been the standard technology to capture and remove CO_2 from point source exhaust gas. However, due to amine's viscosity and corrosiveness, amine has to be used in solution usually below 50 wt%. Since the temperature difference in the absorber and the regenerator is what drives the CO_2 to separate from the amine, this results in considerable amount of energy input to the water (sensible and latent heat) that accounts for more than half of the solution. A different approach is required to dramatically reduce the utility consumption. Our approach is to impregnate porous material with amine, which enables to avoid the use of water solution while gaining the benefit of using amine as the capture agent. The amine is being developed especially tailored to peak its absorption capacity at CO_2 partial pressure 4-5 kPa, which combined with the large surface area of porous carrier, enables the heat of absorption /desorption to be lowered.

From the process side of view, since the CO_2 content in the flue gas is dilute, the energy requirement per unit CO_2 removed for flue gas blower/fan is considerable, thus we have looked into low pressure loss structure of the absorber/desorber to reducing the electricity consumption of the process.

In the presentation, development activity regarding the solid sorbent and post combustion capture process will be discussed along with a project overview.

135c Separation of Uranium from Seawater

Sylvester Gyasi

Technically, it is possible to separate uranium from seawater. This design paper seeks to propose a process intensified way to separate uranium from seawater using amidoxime containing polymer due to the high affinity of the uranyl–carbonate complex towards the polyamidoxime resin. The process justification was based on the economic criteria: production cost, physicochemical criteria: concentration and reactivity, technical criteria: extracting methods from a complex aqueous system. The separation process firstly is the adsorption of uranyl ions using ionic exchange resin in the form of a chelating polymer adsorbent. A buoyant platform with adsorbent polymer made of polyethylene fibres that comes into contact directly with sea water powered by a wind turbine which is mounted on top of the platform, continuously takes the adsorbent from the seawater after campaign time through an elution and regeneration process and then returns it to the seawater for reuse. After the adsorbent has been loaded with uranyl ions, additional downstream processes is explored to recover the uranium including cleansing of the adsorption polymer to remove organic materials, desorption: elution of the adsorbed uranium ions from the adsorption polymer with a suitable solution, purification of the eluent: removal of other desorbed compounds, concentrating the solution, solvent extraction of uranium from the solution with a mixture of an organic solvent and a specific complexing agent and finally conversion into yellow cake or uranium oxide. Next, factors that goes into the Uranium separation cost and financial justifications have been included. In future, we recommend researchers to make use of organically sourced, biodegradable substrates such as chitin which could be a good source of green chemistry alternative to conventional plastic polymers that represents a potential source of an uneasy degradable solid pollutant.

171a Analysis of Wind Speed: A Fourier Neural Network Approach

Abdullah Al-Aboosi, AJ Munoz-Vazquez, Fadhil Al-Aboosi, Mahmoud El-Halwagi, Wei Zhan

This work is motivated by the urgent need to transition towards renewable energy resources, the use of which is preponderant to mitigate environmental and climate threats that result from the growing energy demand and continued dependence on fossil fuels.

The primary focus on wind speed is recognized for its highly efficient and environmentally friendly characteristics, as the efficient generation of electricity through wind turbines is significantly influenced by windspeed, making its accurate prediction a vital component in the design process. However, the existing lack of comprehensive and precise wind speed data brings a formidable challenge from the engineering perspective. To address this challenge, an innovative solution is presented in this work, which considers a Fourier neural network, proving to accurately analyze and predict wind speed data, to an acceptable extent.

The proposed method proves to be advantageous, especially when a windspeed sensor fails at a specific time frame, as it can generate a portion of the missing information by using the previous measurements of windspeed. This approach demonstrates its potential to enhance the reliability and accuracy of wind speed prediction, which is essential for effective energy generation and infrastructure planning.

In this study, a thorough validation of the proposed technique is carried out by meticulously comparing real-world windspeed data with predictions generated by the Fourier neural network. The results affirm the effectiveness and applicability of the proposed approach, reinforcing its potential to revolutionize windspeed data prediction, and advance the utilization of renewable energy sources to fight climate change.

171b A Deep Neural Network Regression of the Cahn–Hilliard Single-Particle Thermal Model for $LiFePO_4$ Batteries

Roger Painter, Ranganathan Parthasarathy, Irucka Embry, Lonnie Sharpe, S. K. Hargrove

Lithium-ion batteries serve as the primary sources of power for electric vehicles (EVs) and hybrid electric vehicles (HEVs). For vehicle applications, battery management systems (BMSs) are necessary to protect lithium-ion batteries from overheating and to ensure optimum vehicle performance. Our approach to developing a BMS was based on recent advances in the application of phase field models for lithium-ion batteries. In particular, our reduced-order model (ROM) utilized a dataset generated from the COMSOL® Multiphysics simulation of the Cahn–Hilliard equation for a single particle of a lithium iron phosphate (LiFePO4) cathode: an example of using a reduced-order model (ROM) based on a single-particle model (SPM). The main innovation of our ROM is that the SPM is fully coupled to a heat transfer model at the battery cell level. We utilized principal component analysis to identify a lower-order model that could reproduce the battery's voltage and temperature response for ambient temperatures ranging from 253 to 298 K and for discharge rates ranging from 1 C to 20.5 C. The reduced-order dataset was then fitted to the experimental data for an A123 Systems 26650 2.3 Ah cylindrical battery using deep neural network (DNN) regression. The entire BMS is realized in conjunction with a digital-twin (DT) configuration with an offboard COMSOL® Multiphysics SPM simulation and training of the DNN, allowing the ROM to be periodically updated by retraining the DNN for aging batteries and actual operating conditions. In this configuration, only the trained DNN predictor function is onboard and in real time.

171c First-Principles Modelling of Photocatalytic Activation of CO2: Challenges and Approaches

Žan Kovačič, Andraz Pavlisic, Petar Djinović, Kristijan Lorber, Blaž Likozar, Matej Huš

Introduction

Titanium dioxide (TiO_2) is one of the most studied semiconductor materials, both experimentally and theoretically, on account of its photoactivity and a range of other potential applications. Recently, its utilization in the field of photocatalysis has become increasingly important as it represents a way to carry out catalytic reactions with lower energy consumption.

Nevertheless, pristine anatase and rutile TiO_2 have some drawbacks, such as high band gaps (above 3 eV), which means that only UV light is able to excite the electrons from the valence band (VB) to the conduction band (CB). In order to boost the photoactivity in the visible light range, modifications which decrease the band gap (utilizing visible light) are required. Moreover, other adverse phenomena must also be suppressed, such as charge recombination, while charge transfer should be increased. There are several strategies to solve these problems: doping, the deposition of co-catalysts, the integration of defects (e.g. oxygen vacancies) or using the so called Z-scheme photocatalyst.

Furthermore, in undoped TiO_2, electrons would sink into the bulk. Doping makes the material catalytically active by allowing electrons to reach the surface. In addition, metal doping controls the selectivity of the catalysts toward the desired products, which can range from CO and CH_3OH to higher hydrocarbons in the case of CO_2 photoreduction.

Since experimental testing of all possible changes is unwieldy, *in-silico* calculations can be performed instead. Increasing computational power has led to an increase in theoretical studies of photocatalysis in recent years. However, the vast majority of theoretical studies focus either on the calculation of catalyst properties or on phenomenological kinetic modelling of experimental results. When a reaction mechanism is studied *ab initio*, including all possible intermediates, activation barriers, and kinetic parameters, calculations are usually performed in the ground state. However, the explicit consideration of the excited state is crucial for the correct representation of a photocatalytic process.

An alternative approach to modelling photocatalytic reactions uses the fact that the dielectric constant of TiO2 is large enough to treat electrons and holes as free as opposed to excitons. In this approximation, the mechanism can also be investigated using electro-chemical methods (grand canonical DFT or JDFT).

Methods

We investigated a simplified mechanism of CO_2 photoreduction on Cu (111), pristine anatase (101), pristine rutile (110) and Cu/rutile (110) and Cu/anatase (101) surface. First, ground-state calculations of the reaction mechanism are performed at the DFT level. Second, excited states are accounted for following Kasha's rule by exciting one HOMO electron from the valence band to the LUMO orbital of the conduction band. The reaction mechanism and kinetics are then compared between different surfaces in the ground and excited state.

Density functional theory (DFT) calculations were performed with the VASP-5.4.1 package. The PBE functional was used as implemented in the PAW pseudopotentials. The energy cut-off of 500 eV was shown to suffice for well converged results based. To account for strong electron localization, the Hubbard correction of (*U-J*) 4 eV was applied to the Ti *3d* electrons in the approach by Dudarev. Long range dispersion forces were considered with the D3 correction by Grimme. Spin polarization was included for all the calculations.

To study surface phenomena, surface slabs were constructed as follows. For Cu (111), a 3-layer 6x6 supercell was prepared. A 3-layer 3x2 supercell of pristine anatase (101) and Cu/TiO_2 anatase (101) surface with one fixed layer, and a 3-layer 3x3 supercell of pristine rutile (110) and Cu/rutile (110) with two fixed layers were constructed. In all instances, 12 Å vacuum was added to the slabs to avoid spurious inter-cell interactions. A dipole correction was applied in the z-direction. The supercells were sampled with a Monkhorst-Pack mesh, at 1x1x1 (Gamma point) due to the size of the supercells. A cluster of four pre-optimized Cu atoms was deposited and allowed to relax on pristine anatase (101) and rutile (110) surfaces. All structures were relaxed until the atomic force dropped below 0.05 eV/Å. The nudged elastic band in conjunction with the climbing image was used to identify the transition states, which were refined with the dimer method.

For the excited-state calculations, we applied the delta self-consistent field (ΔSCF) approach. Firstly the structures of the initial, final and transition states were optimized in the ground state. Subsequently, the density of states was analyzed for each structure to identify the orbital for electron insertion. Then, an electron was moved from the valence band maxima (VBM) into the conduction band minima (CBM), leaving a hole in the VBM. A triplet spin state was used to force the population of the LUMO and depopulation of HOMO. After the change in orbital population, orbitals are reoptimized, and the energy is computed from these orbitals.

To evaluate the time evolution of the heterogeneous photocatalytic process, we created a microkinetic model using DFT-calculated reaction parameters. The model assumes an ideal batch reactor with the following: constant temperature and pressure, no mass transfer limitations, ideal mixing, no lateral interactions. Since the microkinetic model was set up to compare the *differences* in reaction kinetics in the photoactivated and dark regime, the exact reactor parameters was less of a concern. To mathematically describe the transformation, we constructed and solved a system of differential equations for the gas and surface reactions along with a balance for the active sites.

Results

We explicitly considered the excited states during CO_2 photoreduction with the ΔSCF method. Due to Kasha's rule, we only considered the first excited state. For Cu (111), no differences were observed on account of its metallicity. Consequently, the rutile and anatase crystal phases of TiO_2 were chosen as benchmark and the most researched photocatalysts in the literature. We showed that photo-excitation has a significant impact on the activation barriers, which can be significantly reduced. Nevertheless the pristine facets are rather inactive for the CO_2 reduction. Lastly, we deposited four Cu atoms on the rutile (110) and anatase (101) surfaces as a computationally tractable model of the Cu/TiO_2 formulation.

On Cu (111), surface we can expect the direct dissociation of CO_2 to form CO and hydrogenation of CO_2 under *thermocatalytic* conditions. The relatively high CO adsorption energy might hints that it is unlikely to desorb and indeed Cu is a well-known catalyst for the formation of C_{2+} products. There is little change in its photocatalytic performance due to the overlapping valence and conduction bands (no band gap).

On the other hand, pristine anatase (101) exhibits a high band gap of 2.36 eV. The ground state calculations reveal CO_2 would be reduced through direct hydrogenation, but this is unlikely due to a high activation barrier. Inclusion of the excited states significantly reduces the barriers. CO adsorption strength is rather weak, which is also the case for pristine rutile (110) surface. A direct CO_2 dissociation is again impossible due to unfavourable thermodynamics. The most likely reaction pathway is the direct hydrogenation to HOCO, which then likely reacts further towards methane or methanol. The reaction rates in all cases are negligible.

Finally, we deposited four Cu atoms on rutile and anatase. Contrary to pristine anatase (101) surface, the Cu/anatase (101) surface has much lower activation energies in the ground state, where results suggest that hydrogenation and dissociation of CO_2 are almost equally likely to happen. After the formation, HOCO can then react further or dissociate to CO and OH. Inclusion of excited states furthermore reveals that the activation barriers are further lowered. In the case of Cu/rutile (110), it was found that in the ground state the direct dissociation of CO_2 is likely the predominant pathway, whereas hydrogenation has a higher barrier. Due to low adsorption energies of CO, it is also likely to be the major product. The inclusion of excited states even further lowers the E_A.

It should be noted that Cu-clusters on a TiO2 surface introduce cluster states in the band gap, which themselves do not significantly influence it as shown in Figure 2. This is expected as cluster states do not have a sufficient cross section to influence an optical measurement but they can function as recombination sites. Consequently, the photo-excited electrons have substantial energy inside TiO2 but the overall efficiency of the system can be lowered by the recombination sites.

Conclusion

Subsequently, a microkinetic model was constructed to investigate the evolution of the participating species on all five photocatalytic surfaces. In the ground state, CO_2 reduction did not proceed at the mild reaction conditions of 1 bar and 298 K, which was expected. However, accounting for the excited states revealed that Cu/rutile (110) and Cu/anatase (101) surfaces are active. The former was active for water splitting, forming predominantely H_2, whereas Cu/anatase (101) favored the production of CO. The results were found to agree with other experimental studies, reinforcing the importance of including excited states in DFT calculations and subsequent macro-scale models.

AUTHOR INDEX

AIChE
120 Wall Street, FL 23
New York, NY 10005-4020

CURRAN ASSOCIATES INC.
proceedings.com

ISBN 979-8-3313-1714-0